8/97

Sherwood

Hope you enjoy

the book!

Best Regards

Danny

POLICY EVOLUTION:

Energy Conservation
to
Energy Efficiency

A Series of Speeches
from the
Energy Efficiency Forums

POLICY EVOLUTION:

Energy Conservation
to
Energy Efficiency

A Series of Speeches
from the
Energy Efficiency Forums

Compiled by
Douglas A. Decker, P.E.
with
Alan Berolzheimer, Ph.D.

Published by
THE FAIRMONT PRESS, INC.
700 Indian Trail
Lilburn, GA 30247

Library of Congress Cataloging-in-Publication Data

Descker, Douglas A.
Policy evolution : energy conservation to energy efficiency : a series of speeches from the energy efficiency forums / compiled by Douglas A. Decker with Alan Berolzheimer.
p. cm.
Includes index.
ISBN 0-88173-274-5
1. Energy policy--United States--Congresses. 2. Energy conservation--United States--Congresses. 3. Energy consumption--United States--Congresses. I. Berolzheimer, Alan R. II. Energy Conservation Forum. III. Title.
HD9502.U52D444 1997 333.7'0973--dc21 97-12267
CIP

Policy Evolution : Energy Conservation To Energy Efficiency : A Series Of Speeches From The Energy Efficiency Forums By Douglas A. Decker with Alan Berolzheimer.

Published by The Fairmont Press, Inc.
700 Indian Trail
Lilburn, GA 30247

Printed in the United States of America

10 9 8 7 6 5 4 3 2 1

ISBN 0-88173-274-5 FP

ISBN 0-13-673716-1 PH

Distributed by Prentice Hall PTR
Prentice-Hall, Inc.
A Simon & Schuster Company
Upper Saddle River, NJ 07458

Prentice-Hall International (UK) Limited, London
Prentice-Hall of Australia Pty. Limited, Sydney
Prentice-Hall Canada Inc., Toronto
Prentice-Hall Hispanoamericana, S.A., Mexico
Prentice-Hall of India Private Limited, New Delhi
Prentice-Hall of Japan, Inc., Tokyo
Simon & Schuster Asia Pte. Ltd., Singapore
Editora Prentice-Hall do Brasil, Ltda., Rio de Janeiro

Table of Contents

Preface

In the course of one generation, profound questions about energy and its use have become prominent in public consciousness in the U.S. as never before. Spurred by dependence upon imported oil, questions about nuclear power, environmental degradation and resource depletion, Americans today maintain interest in energy policy and energy politics.

To what degree of interest depends on the events of the day. At the first forum, Rep. Les Aspin (D-Wis.) pointed out the national security concerns surrounding energy conservation, despite the fact that federal spending on energy was extremely low. Then, during the war in the Persian Gulf, there was a great deal of preoccupation with energy policy. But during peace, prosperity and low oil prices, the public's interest wanes. The key is to go beyond the swinging pendulum described in 1991 by former Energy Secretary James Schlesinger and to move to a sustainable energy ethic that overrides political and economic fluctuations.

As our public discussions about energy politics and policy have become more comprehensive and sophisticated during the past two decades, the concepts of conservation and efficiency have assumed critical significance. The subtle yet important distinctions between these two concepts illuminate how public discourse about energy has evolved, and they highlight the path along which many informed observers of the energy scene expect energy policy to travel in the foreseeable future.

From the Beginning

This book explores the transition from conservation to efficiency as reflected in the proceedings of a series of annual conferences on energy that began in 1990. That year Johnson Controls, Inc., a leader in the energy management industry, convened a national forum of participants from the public and private sectors as part of a broader campaign to raise energy conservation awareness in Washington, D.C., and to encourage the development of a nationwide energy conservation ethic.

In 1992 the United States Energy Association, an association of nearly 200 public and private energy-related organizations and the U.S. representative to the World Energy Council (WEC), joined Johnson Controls as a cosponsor of the forum. Over the course of these seven years the discussions and debates at the annual forum have exhibited both continuity and change.

Questions about the role of government in the marketplace, public/private partnerships, promoting research and development of energy-efficient technologies, market opportunities for U.S. businesses, the global context, and the critical impact of energy use on the natural environment have been a consistent thread. Changes in the political landscape have influenced the character and tone of the forum from year to year, occasionally in dramatic ways.

Most important, and central to the purpose of this anthology, the record presented by these forums shows how the orientation of energy experts in government, private industry, and the non-profit policy research sector has shifted from an emphasis on conservation to an emphasis on efficiency.

Energy Conservation Initiative

Johnson Controls sponsored the first annual Energy Conservation Forum as one component of its Federal Energy Conservation Initiative, which also included presenting testimony before Congress, discussions with congressional staffs and federal agency managers, an extensive media campaign, and the formation of a coalition of leading business and construction industry organizations to encourage and support President George Bush's decision to sign an executive order on federal energy conservation.

In the early 1990s, Johnson Controls and the United States Energy Association sought to instill an energy conservation ethic in the U.S., among citizens and especially among policy-makers. Much of the momentum for energy awareness generated by events of the 1970s, such as the 1973 oil embargo and the 1979 accident at Three Mile Island, had dissipated during the 1980s.

One reason for this stems from President Ronald Reagan's orientation to energy policy. During the Reagan administrations, the

U.S. government took strenuous action to ensure America's access to oil on the one hand, and on the other hand drastically scaled back federal spending on other energy-related initiatives, including renewable technologies and the encouragement of investment in efficiency measures.

The tide began to turn with George Bush's executive order on federal energy management. When Johnson Controls inaugurated these annual forums, its primary objective was to reinvigorate the federal commitment to promoting energy conservation. The first couple of forums focused heavily on implementing conservation and efficiency measures within the federal government itself, the largest consumer of energy in the entire world. One of Johnson Controls' and United States Energy Association's primary goals, shared by many energy managers in the federal sector, was to make the federal government an outstanding model of energy conservation for the American public.

Conservation Gives Way to Efficiency

As the forum expanded, as the political climate in Washington fluctuated, and as concern about federal agency budgets intensified, the ethic of conservation at the root of the original campaign gave way to a more diversified conceptualization of energy efficiency.

Many conference participants expressed the conviction that "conservation" implied not using, doing without, and diminished expectations. While there was a strong consensus about the enormous value of conserving resources and using them judiciously, the sentiment prevailed that the fundamental goal should be the wise use of energy. For the federal government, that meant helping agencies reduce costs, be more productive and fulfill missions. For the general public, it meant economic and employment growth, environmental protection, equitable sustainable development and broad-based prosperity. Where "conservation" suggested stagnation, the concept of "efficiency" was dynamic.

The concept of efficiency encompassed a wide range of ideas and actions designed to use energy effectively in the pursuit of human progress. In 1994, this reorientation was formalized through a change in the name of the conference: it became the Annual Energy Efficiency Forum.

New Focus

Reflecting the structural dynamics in the U.S. economy and the energy industries in particular, forum participants paid increasing attention to issues such as research and development on new technology, domestic and international markets for U.S. efficiency goods and services, financing the implementation of efficiency measures, and the brave new world of utility deregulation and competition.

At the time Johnson Controls Vice President Brian Stark noted, "Four years ago we called attention to a serious problem. Today there is a consensus that the problem exists. Now we need to build on our understanding of the problem and help find solutions to carry us into the next century."

Of course, no one lost sight of the crucial question of the role of government, especially in the wake of the startling ascendancy in Congress of Republicans passionately committed to reducing the size and scope of the federal government.

In 1990, the big questions centered on how to encourage a conservation ethic among Americans and promote energy efficiency within the government apparatus. By 1996, the big questions centered on how to preserve and enhance the delivery of energy efficiency to residential, commercial, and industrial consumers under new and rapidly changing market conditions, to ensure that the social, economic, and environmental benefits of wise energy use continue to be realized.

Forum Objectives

DOE's Energy Information Administration's 1997 Annual Energy Outlook projects that rapid technology development can lower energy production costs and boost production over the next 20 years. The resulting lower energy prices combined with projections for national economic growth could mean, once again, a diminishing interest in energy efficiency. However, the Outlook also notes that rapid penetration of efficient technologies in end-use markets can slow growth in energy demand, right on target with the objectives of the forums.

By making energy efficiency a top bipartisan priority among policy-makers in Washington, encouraging federal investment in energy efficiency measures, and creating a strong business/government alliance to ensure the successful implementation of conservation-oriented policies, the attendees of the forum can continue to have a major impact on environmental problems, the U.S. budget deficit, global competitiveness, oil imports, and national security.

The editors have chosen the selections for this book with the objective of providing the reader with an insider's view of the evolution of energy policy attitudes from the ethic of conservation to the ethic of efficiency. Along the way, we have sought to fairly represent the variety of themes and perspectives presented at the forums, and to capture the prevailing mood and tenor of the proceedings.

In some ways, a great deal has changed since 1989. The opinions of some speakers may be different now due to modifications in the marketplace, political climate, technology and the impending deregulation of utilities. In other ways, not much has changed at all: energy efficiency remains an excellent way to reduce pollution, create jobs, save energy and save money.

We hope this synopsis will itself contribute to the ongoing policy evolution, and advance the ideals of energy efficiency throughout the world.

Douglas A. Decker, P.E.
Alan R. Berolzheimer, Ph.D.

ACKNOWLEDGMENTS

This book is dedicated to all those who understand the need for energy efficiency in today's world and promote it for tomorrow.

In a spirit of bipartisanship, we acknowledge the personal contributions to the causes of integrity and leadership from the late Rep. Sherman P. Lloyd (R-Utah) and Rep. Mike Synar (D-Oklahoma).

The authors wish to thank all of the speakers at the forums throughout the years for giving their time and talent toward increasing information about the vital field of energy efficiency.

We would like to recognize the Forum Planning Committee for their endless hours of work in guiding the philosophy and theme of each event, including John Armstrong, Henri-Claude Bailly, Mark Hopkins, John Millhone and Barry Worthington.

We also would like to thank the dozens of people who have worked behind the scenes and in front of the registration table, including David Beightol, John Bernaden, Barbara Haig, Michelle Jones, Alison Klukas, Linda Montalvo, Beth Monroney, Gail Niedernhofer, Annmarie Pittman, Glen Ponczak, Carol Rowland, Richard Spees, Noelle Tower and Mark Wagner.

Additionally, we would like to honor the members of the United States Energy Association Board of Directors who provide their support to the event, especially P.J. Adam and Don Jordan.

We also would like to thank those at Johnson Controls, Inc. who have backed this event from day one, including James H. Keyes, Joseph W. Lewis, Brian J. Stark, and Thomas M. Gannon.

Finally, we wish to thank those who assisted in the production of this book, especially Barbara Haig and Linda Montalvo.

ABOUT THE AUTHORS

Douglas A. Decker, P.E.
Vice President, Government Business Unit
Johnson Controls, Inc.

Mr. Decker has been actively involved with energy issues throughout his career. He has testified in a number of congressional hearings and contributed to government studies on the benefits of energy conservation and efficiency in federal buildings. He has been interviewed frequently in the media and has authored numerous articles on energy efficiency issues. His latest work is included in *Engineered Systems* and the *Encyclopedia of Energy Technology and the Environment*, published by Wiley-Interscience.

A registered Professional Engineer in Wisconsin and California, Mr. Decker is a member of the National Society of Professional Engineers and was honored with a Special Recognition Award from the Association of Energy Engineers in 1995. He is a frequent lecturer to national audiences.

Mr. Decker is a graduate of the University of Denver, with a B.S. degree in Business Administration and Engineering.

Alan R. Berolzheimer
Editor

Mr. Berolzheimer has edited and contributed to the production of a number of books for several publishing firms, including Chelsea Green Publishing, Terra Nova Press, Jewish Lights Publishing and Steerforth Press. A former editor of The Papers of Daniel Webster, he also has written papers on the electric utility industry.

Mr. Berolzheimer graduated cum laude with a B.A. in History from Dartmouth College, and received his M.A. and Ph.D. in U.S. History from the University of Virginia.

First Annual

Energy Conservation Forum 1990

First Annual Forum - 1990

In May 1990, Johnson Controls, Inc. inaugurated the Energy Conservation Forum as part of a public information campaign designed to raise awareness in Washington, D.C. and to encourage the development of a nationwide energy conservation ethic. This campaign focused on the federal government, because it is the largest user of energy in the U.S. economy, with the goal of making the federal government an outstanding model of energy conservation for the American public.

The first forum was a low-key event on Capitol Hill attended by several congressional members, staffers, agency representatives and reporters. No recordings were made of the proceedings, but media articles pointed to the importance of the message. Then-Congressman and chairman of the House Armed Services Committee Les Aspin spoke to the attendees and highlighted the magnitude of energy efficiency as a national security issue.

"One of the aspects of energy conservation that hasn't received much attention is its effect on national security," Aspin said. "After the energy crunch of the 1970s, we reduced our dependence on foreign oil. But over the past ten years, our dependence has crept up towards the pre-energy crisis level. We haven't heard much about this as a political issue, but if we get hit with another energy crunch, you can bet that this issue will come back in spades," he said.

Aspin said that with the unfolding events in the Soviet Union and Eastern Europe, the definition of national security was changing to include new aspects such as energy independence, adding that "energy conservation is a preventative medicine for our energy security." Ironically, just eight months later war broke out in the Persian Gulf.

At the forum, it was noted that while the federal government was the largest single consumer of energy in the world, it spent only an average of $121 per building on funding energy conservation retrofit and capital expenditures in 1988. Overall energy conservation funding for federal agencies declined by 80 percent between 1981 and 1988 while expenditures for energy increased by $2.2 billion.

Additionally, Department of Defense figures indicated that it spent about 81% of the government's total $3.7 billion annual energy bill, and that it actually deleted all expenditures for energy conservation for 1989.

Aspin also noted that he had written a letter to U.S. Secretary of Defense Richard B. Cheney urging a re-evaluation of DOD's policy on energy conservation and adding that the House Armed Services Committee and 21 U.S. Senators supported his position. Aspin said that the DOD's role as the federal government's largest consumer of energy gave it an opportunity to set an example for other government agencies.

The enthusiastic response to Aspin's words and the entire forum inspired Johnson Controls to continue the event, and work began on the next forum almost immediately.

Les Aspin, former Secretary of Defense and Congressman from Wisconsin, spoke about energy efficiency as a matter of national security at the first Energy Efficiency Forum in 1990.

Douglas A. Decker, P.E., Vice President, Government Business Unit, Johnson Controls, Inc. founded the Energy Efficiency Forum in 1990 and serves as master of ceremonies.

Second Annual
Energy Conservation Forum
1991

Second Annual Forum – 1991

The Second Annual Energy Conservation Forum expanded from the loosely structured first event to a longer, more formal event. Operation Desert Storm, America's powerful military foray into the Mid-East, was still fresh on the minds of the participants as once again, the world focused on the urgent need to keep a steady oil supply.

Dr. James R. Schlesinger, who in his career had served as chairman of the Atomic Energy Commission, director of the Central Intelligence Agency, Secretary of Defense, and Secretary of Energy, delivered a keynote address that focused on the interconnections between energy use, energy conservation, politics, and national security. His dry honesty startled many in the audience, just as it will startle some readers: "It's a hell of a lot easier and it's a hell of a lot more fun to kick ass in the Middle East than it is to actually make sacrifices in order to reduce the importation of oil," he said.

Dr. Schlesinger's speech was followed by a panel discussion about the funding and implementation of the Executive Order on Federal Energy Management signed by President Bush in February 1991. The panelists included members of Congress and employees of both the executive and legislative branches of the federal government knowledgeable about energy issues.

While they all agreed about the importance of federal energy conservation, the panelists disagreed about the significance of what had been accomplished to date. They shared a belief that the federal government had a long way to go before it could claim to stand as an energy conservation model for the rest of the country. We present the comments of four of the panelists.

ENERGY CONSERVATION: POLITICS, SECURITY, AND THE ENVIRONMENT

James R. Schlesinger served as the first Secretary of the U.S. Department of Energy, from 1977 to 1979. Prior to that, he served as Secretary of Defense, Director of the Central Intelligence Agency, and Chairman of the Atomic Energy Commission.

James R. Schlesinger: What I'm going to do today, is to look at energy efficiency and energy conservation in a somewhat broader context. There are some engineers in this group that may not realize it, but by itself and without that larger context, energy efficiency does not always have a great deal of public appeal. So I shall attempt to provide that larger context.

The first point that I want to make concerns the transformation of the world scene that has occurred in the last two years – a change that has now become a cliché. The end of the Cold War, more recently the successful war in the Gulf, has altered the international scene in ways that none of us could anticipate. But those changes are also changing energy policy. As a result, energy policy is altered from what it has been historically. More precisely, energy policy is different from the way it has generally been interpreted in the period since the Arab oil embargo. That change has been reinforced by recent events in the Persian Gulf.

Some of you may have heard the rumor that our involvement in the Persian Gulf, starting in the summer of 1990, may have had something to do with the energy supply. Some of our leaders stated that candidly in August and September of 1990. The public reaction was: we are not going to fight a war in order to hold down the price of gasoline. And so, the public rationale was adjusted. But the original view was correct: the events of the Gulf did flow from the issue of energy supply. As a result of the war, there has been a dramatic change in the balance of power in the Persian Gulf itself, that has made our energy supplies much less at risk than what had previously been the case.

In considering energy policy, in this broader context, I'm going to start with a bit of history. Until 1945, most of America's energy consumption was coal. One does not have to be reminded that coal was difficult to handle – dirty, costly. After 1945, when the Middle Eastern oil fields opened up, the industrial nations, perhaps most notably, Western Europe shifted from coal to oil. They largely gave up using coal and began to use oil to such an extent that consumption of oil worldwide grew at a rate in excess of seven percent per year, indeed, more than doubled in each decade. Here in the United States there was an enormous growth in the consumption of petroleum – which reflected a conviction on the part of the average American citizen that the Lord God deemed that he had a right to plenty of gasoline at a low price.

That momentum ceased suddenly in 1973 with the oil embargo. It was at that juncture that we recognized our economic and political vulnerability to events taking place in the outside world, including a possible interruption of our oil supply. For those of you who recall the details of the 1973 crisis, the Soviet Union was deeply involved. The Soviets went around from country to country in the Middle East urging those nations to join in the oil embargo against the West. So the events of 1973 changed our national attitudes. They underscored both the physical vulnerability of the United States on national security grounds (particularly since the Soviet Union lay so close to the Persian/Arabian oil fields), and it also underscored our economic vulnerability.

After the 1973 oil shock and the subsequent oil shock in 1979, we had a bout of inflation, followed by a recession and substantial unemployment. But in 1973 the political system reacted. The political system of the United States is enormously sensitive, believe it or not, to the cries of the public. And when the public believes something to be important, there will be an immediate response on the part of political figures. Indeed, after the Arab oil embargo, there was a flood of legislation. Some of it was good, a lot of it was just simply wild. But it responded to the public demands. And, particularly, public demands that something be done about the energy crisis.

The administrations involved after 1973 also reacted in the same way. Never again! President Nixon launched his Project Independence. President Ford's energy policy also stressed diminished dependence on foreign sources of supply, and President Carter's national energy policy made as a centerpiece the reduction of our importation of oil. In President Carter's National Energy Plan (NEP), you will recall, conservation was the cornerstone. There was some derision, a subject that I will return to later on. But continuing with the history, the pattern was broken temporarily with President Reagan. President Reagan believed in the market – the magic of the marketplace. Indeed, he believed exclusively in the magic of the marketplace. And he did not think that there was an energy problem, if only we could get the government off our backs. He asserted that there was more oil in Alaska then there was in Saudi Arabia, I perfectly hope that proves to be true. He tended to scoff at conservation. He said conservation was when one was warm in summer and cold in winter. But with President Bush, the concern about energy use has returned, and energy policy has revived. That energy policy has revived was reflected, for example, in the Executive Order of last April, which instructed the various agencies and departments of the executive branch to get with the program of energy conservation.

But understand this: the problem of energy policy in the context of 1991 is radically different from the problem that we saw in the period after 1973. The end of the Cold War and the events in the Gulf in the last year have altered the emphasis in energy policy. One important element: the Soviet threat is now substantially abated. For many years we were concerned, rightly concerned, that the Soviet Union was close to the oil fields of the Middle East and might move in and seize control. If the Soviets had control of the oil tap in the Middle East, the international order, as we had shaped it after the Second World War, would come to an end. The Japanese and the Europeans would have been forced to make an accommodation with Moscow. Soviet military power was close at hand, their airborne divisions, unfortunately, lay to the north of those oil fields. But that threat has now gone away to a large extent. And our capacity to deal with any such threat has grown. The Soviet Union, or what used to the Soviet Union, is going to be preoccupied with domestic reconstruction and will not be in a position to indulge in foreign adventures. So one aspect of the national security problem is much abated.

Our second concern was about hostile rulers in the Middle East. That started with Colonel Nasser, but was strongly reinforced when Colonel Kaddafi overthrew the monarchy in Libya and began to play around with the oil issue. It was further reinforced when the Shah was overthrown, and the Ayatollah Khomeini shut down Iran's oil fields, and sought to bring punishment to bear on the Great Satan. In that period, Saddam Hussein was our chosen counterweight to the Ayatollah Khomeini. Subsequently, as you know, Saddam Hussein got out of hand, and was taken down with great speed and effectiveness. And as a result, the fear that we had for many years of some regional leader attempting either to defy the United States or to dominate the oil fields strategically is also much abated.

How does this affect energy policy? When Saddam Hussein seized the oil fields of Kuwait and threatened those in Saudi Arabia, oil prices began to rise. There was an immediate public reaction, just as there had been in 1973 and 1979. Concern about rising prices – which is what tends to drive the public – led once again to the usual political rhetoric. A bundle of legislation was thrown in the congressional hopper. But the success of our military ventures swept that away. With the start of Desert Storm, the oil price fell by $10 per barrel.

Public concern began to wane. The moral that the U.S. public drew from the events of the Middle East was quite simple. It's a hell-of-a-lot easier, and it's a hell-of-a-lot more fun, to "kick ass in the Middle East" than it is to actually make sacrifices in order to reduce the importation of oil. (I apologize for any words I might have used in these last sentences, but they are not mine. I draw them from the highest authority in the land.)

With the drop in oil prices and with the change in public attitudes as a result of the quick victory in the war, the thrust behind energy policy once again diminished. No surprise, it's happened before. So what we have seen as a consequence is a transformation, not only of international politics, but a transformation of the priorities in energy policy. **After 1973 the new age had the following priorities: first, national security; second, reflecting the public's attitude, low-cost energy; and third, the environment.**

Environmentalists tried to take advantage of the new concern about energy to move along environmental objectives. But those objectives were only third in the priority list, and they were only trying to creep in. A number of our responses to 1973 were not environmentally friendly, most notably the desire to convert to coal. Coal was then described as "America's ace-in-the-hole." If you have been watching the environmental legislation of late, you would recognize that is not entirely what is currently in mind in terms of the Amendments to the Clean Air Act.

Those three priorities are now probably reversed. Thus, from the public standpoint, the chief priority is the environment; second, low-cost energy; and third, national security. All this reflects, of course, the changes in the international environment. The most serious legislation impacting energy over the last decade is not a piece of energy legislation. It is the Clean Air Act, which will affect energy in many ways. The priority on the environment reflects strong public support. The public is seriously concerned about air quality. It is concerned about air quality in national parks – the haze that hangs over the Grand Canyon. In New England there is concern about acid rain. In southern California, the Southern California Air Quality Control District is now contemplating restrictions on the use of the automobile and on backyard barbecues. This is in California! When such things occur, you can conclude that something is afoot in terms of public opinion. A change in public attitude does not necessarily mean a change in public decisions. In addition to the concern about our air quality, there also is concern about the greenhouse effect. The European Community, for example, is now considering a carbon tax as a way to raise money. There is general awareness that the use of energy is the dominant source of pollution.

The other priorities have not gone away. After the environment, low-cost energy continues to be the public's principal concern. That has been reinforced by the current economic climate and by the decline in real income per capita among people in our middle-income groups. However, there is, serendipitously, an enhanced discovery rate and improved technology expanding natural gas supplies. The enlarged supply of natural gas can simultaneously hold down energy costs for the public while reducing the level of air pollution, and indeed, for those who worry about the greenhouse effect, the amount of carbon dioxide put into the air. National security considerations have not gone away, but they have been reduced. It continues to be recognized that the improvident use of energy, and particularly of oil, will increase our dependence on the Middle East, and that the Middle East continues to be an unstable region.

All of which brings me to the focus of these sessions: conservation and energy efficiency. The critical point to bear in mind is that conservation unites all three of these priorities. It perhaps most notably contributes to what is the fashionable priority of the moment, the environment. Less energy, less oil, means less air pollution, it means reduced impact on possible global warming.

It is not true, as you may have heard, that Mark Twain said, 'Everyone talks about conservation, but nobody does anything about it.' Much has been done over the last 18 years. Something continues to be done. But efforts in terms of energy efficiency and conservation have slowed in recent years. In order to spur energy efficiency, we need inducements and pressures which we presently lack.

Let me cite the case of Japan. Japan, as you know, imports 99% of its oil. It was really frightened in 1973. Its GNP dropped sharply for the first time since its takeoff into sustained growth after World War Two. And then in 1979, when the Middle East and oil fields were once again partially shut, panic really spread in Japan. The conservation of energy became a national preoccupation. With the famed consensus that determines policy in Japan, what was a national preoccupation has resulted in continuous action. Industry was steadily prodded to save energy. It was made quite clear that it was the duty of every Japanese to save energy in whatever way he could. As a result, Japan has become substantially less vulnerable to international disturbances. It rode out the brief run-up in energy prices that occurred last year without much travail. In 1973 the cost of the importation of oil represented by itself some 4 percent of the GNP of Japan. By 1990, that had been reduced from 4 percent to 1 percent. Through improvements in energy and efficiency, the Japanese reduced their vulnerability to events taking place in the outside world – and protected economic growth, which as you know is still the principle objective of Japanese policy.

There is next the case of U.S. business. I refer now to <u>users</u> of energy. U.S. business took note of the rough-shod run-up of prices of energy in the period after 1973. For U.S. business, no further inducement than market forces was needed. American industry has done quite well in terms of increasing the efficiency of energy use. But much energy in the Untied States is not consumed by business; it's consumed by average citizens. U.S. citizens are much more inclined to follow habits, not to keep as sharp a pencil as business concerns, to grumble about, but not to adjust their behavior as a result of change in price. U.S. citizens have to be persuaded to change their habits. Industry responds to market inducements; the ordinary citizen needs some kind of prod in order to move toward greater energy efficiency.

What moves the society toward greater energy efficiency? Three things: first, there is the effect of higher prices. When prices are rising, industry at least, and to some extent the consumer, will use less. That rise in price, if it is reversed, contributes to a slowing-down of the movement toward greater energy efficiency. A rise in price, however, can be reinforced by taxes. Now since the late 70s, it is difficult to use taxes in the United States as a way of reinforcing the price mechanism. First, because of a tax revolt that started in California and is embodied in Proposition 13; and then more recently, because of the budget impasse between the White House and the Congress. I noted the other day that John Dingell said, in place of more regulatory standards, what the United States needed was higher taxes. Today, there are few people in the land who are prepared to say that.

A second instrument is jaw-boning. If market forces and price movements do not move the society toward greater energy efficiency, jaw-boning may help. And setting a good example. If the federal government fails to set a good example, then all of the jaw-boning will be to no avail, because the public will take note of the fact that the federal government tends to be an energy squanderer.

A third tool, much used in the past, less used today, is the direct regulation of, or intervention in the marketplace. I have little scruple about using intervention in the marketplace, if it really would achieve greater efficiency in the use of energy, or greater use of domestically available fuels. But I recognize that this subject is controversial, and I do not wish to enter into a vexed dispute at this time. I merely point out that regulation can increase energy efficiency. Societies other than America have been prepared to use energy regulation as the way to push down energy use per unit of comfort. We have been reluctant to do so – but that is a debate that will likely continue in this country.

Let me conclude with a few observations. First, there is a national consensus with regard to energy efficiency. Energy efficiency is a good thing. Conservation is a good thing. Everyone agrees on that. But conservation means different things to different people. That there now is a national consensus about and support for energy conservation is somewhat novel. When President Carter stated that conservation was the cornerstone of the energy program, there was an adverse reaction. Conservation was not the American way! Production was the American way! In the course of the last 12 years, that kind of attitude has disappeared. It is now universally recognized that conservation, if only for environmental reasons, is an appropriate objective for both the society at large and the federal government. The change has been so great that everyone is now obliged to pay lip-service to energy efficiency. At least lip-service. Even those who sell energy, who are somewhat ambivalent on the subject of energy conservation for reasons that I need not dwell on, continue to pay their respects to energy conservation.

As the country has changed in this period since the TET offensive and the breakdown of our bipartisan consensus on foreign policy, there are few things that are truly bipartisan. Conservation is now one of those things. It earns the support of politicians of every persuasion. Nonetheless, that does not determine policy. The establishment of an agreed-on end, such as energy conservation, does not mean that there will not be quarrels over means. It is not the goal about which there is any quarreling; there is no longer any scoffing about the desirability of energy conservation. But there continues to be, as I mentioned a moment ago, particularly in the area of the degree of regulation, dispute over means. Energy policy is particularly difficult in our society. It reflects all the contradictions in the society. It reflects our diversity as opposed, let's say, to national security objectives. The war in the Gulf tended to unify the country; energy policy tends to divide the country. And with the change in our political system, there are more and more groups that can say no to any particular policy. For that reason, it becomes harder to formulate an energy policy in this country, as President Bush and Secretary Watkins have lately discovered, or rediscovered.

Still, despite all these differences, there is much to be done on which we can all agree in principle. All of us, I think, agree that conservation is desirable when it is cost-efficient. All of us agree that there are many areas in which efficiency can be improved with modest investments of capital. There is no disagreement, I think, that those improvements in energy efficiency that reflect the marketplace are desirable. Given what is at least a minimal level of agreement regarding this issue, it is the national task to get on with the job.

We shall have opportunities after we have achieved progress in the many areas, including the federal government, in which our use of energy can be made far more efficient, cost-effectively, to quarrel about whether we, like the Japanese, should press on with the objective of energy conservation even when there is no financial return in prospect. I stress this because we are likely, if we fail, to have further problems down the line. The Middle East continues to be tempestuous. Even though the Soviet Union has diminished as a threat, there will still be problems in the Middle East – and the possibility, indeed, the likelihood down the line – of further supply interruptions. And we shall have increasingly serious problems with regard to the international balance of payments. But where we are heading at the present time is that by the end of this decade, we will be seeing oil imports at 10 or 11 million barrels per day, and the import bill will likely be well in excess of $100 billion per year. That threatens the balance of payments – even as the United States seeks in the economic realm to demonstrate continued leadership, as it does in the military realm.

Efficient pursuit of energy conservation, and oil conservation, in particular, is necessary if the United States is to play a role in the 21st century as effective as it has played at the close of the 20th century. Thank you.

Moderator: Thank you very much, Dr. Schlesinger. I hope things are really brighter than that, but I guess that sets the tone for the seriousness of our panel discussion today. As the moderator here I guess I get to ask the first question, and my question to you is, is the area of energy and the environment going to be a presidential issue, or will the item just kind of fall off the issue scale in terms of importance in the United States?

Schlesinger: Political issues reflect what the public is interested in. The public's interest in the environmental issue and in the energy issue was at a high peak, something over a year ago. The environmental issue, I think, will continue to be a matter of great concern to the public, in fact one of growing concern, particularly if the hypothesis of global warming turns out to be true. But for the moment, it has receded because of the public's concern with the current recession. The fact of the matter is that the public is preoccupied with economic issues at the present time, understandably so. When the economy revives, we should have once again a revival of interest in the environment and in energy efficiency. That is a luxury we will be able to afford under those circumstances.

Q: Why not bring back the tax cuts, to encourage people to use energy-efficient products in their homes?

Schlesinger: Subsidies is just a reverse of taxation. We can induce the kind of behavior that you want through subsidization, through the tax codes. And, once again, I am a strong believer in that. Why not bring it back? Well, the answer to that is quite simple. It's very hard under the budget rules that have been set up. You have to balance all the reductions and other expenditures, or increases if you're paying taxes, or any subsidization. The budget impasse is a clear impediment to energy policy at the present time. The tools of energy policy, aside from jaw-boning, consist of taxes and intervention in the marketplace. The latter is not embraced for ideological reasons, as one might have noted in the run-up to the national energy strategy when the Department of Energy proposed certain ways of dealing with conservation issues – it was rebuffed by senior members of the White House staff. I think that one of the reasons we have difficulty in formulating a new policy at the present time is that the tools necessary for an energy policy are just not available. One of them is subsidization of conservation measures. I happen to be for that.

Q: How would you assess the progress that America has made in energy conservation and efficiency in the past 20 years?

Schlesinger: Before this era in which sexism became a serious charge, Henny Youngman used to say, "How's your wife? ...Compared to what?" Well, here the question is compared to what? Compared to the Japanese, we have just not done very well, probably we have not done as well as the Germans, the Scandinavians, or the British. On some kind of relative scale, compared to other nations, we still have a lot to accomplish. We started with a big gap, partly because energy costs have always been low, and non-risky in the United States. In Europe, energy has always been a matter of concern, partly for national security reasons, partly because energy prices were much higher. The same is true of Japan. Though if one looks at our present level compared to other countries, we still have a long way to go. You might give us a modest score on that basis. If you look at the improvement that U.S. Business has made since 1973, we have done much better. We have improved as much as the Europeans have improved. But we started from a less advanced place, and still we just reached the same percentage improvement. We have a long way to go.

Now, if you compare U.S. business with U.S. citizens, I think U.S. business has done startlingly well by that standard. Business keeps a sharp pencil, business has noted that there is a substantial rate of return to be made by investing in energy conservation, and has proceeded over the course of the last 18 years in the right direction. The U.S. citizen has been much less sensitive to those considerations. Business has done well, but it still has a long way to go. I'd give it C+ for that.

Q: To continue that question, how has the federal government done in relation to industry?

Schlesinger: The federal government has an even longer way to go, partly because there is no way for an agency or department to see that the results of conservation come to benefit that department or agency. In the case of a business concern, if it invests in energy conservation and cost efficiency, the results will be reflected in the bottom line. That's not the case for a department or agency. And, therefore, what we have had over the period since 1973 is a succession of directives from various presidents, starting with President Nixon, to the department of energy, and other departments and agencies, to get with the energy conservation program. And each year, as that department or agency examines how it will allocate its resources under pressure from the Office of Management and Budget, it decides that the mission objectives of the agency are more important, let's say, than saving energy. Over many years, I had the pleasure of running several departments, agencies, and myself, I must confess, that when I acted trading off mission objectives against conservation objectives, the mission took priority. I suspect that other agencies have taken the same route. And given the set of incentives that exist at the present time, it is not surprising that despite splendid rhetoric from the federal government, its performance has been much poorer than U.S. industry.

Q: How do we develop momentum, promote energy conservation, and will that dilute the chronic problems you discussed, rather than the series of a few acute problems?

Schlesinger: Well, that question is one that we haven't resolved yet. The great benefit of the U.S. political system is it's responsiveness to the public. And if the public has divergent views, the political system will reflect those divergent views. Under a parliamentary system, a government can put through energy conservation objectives and legislation to match. But in the American society it is difficult to do that, and it is difficult to achieve and sustain that momentum. One has got to have an investment of political capital regularly at the highest levels. And given the fact that the highest levels tend to worry about other objectives, it is not an easy task. I don't think that we are likely to achieve the kinds of results that one achieves in other industrial countries. We have had three public reactions, 1973, 1979, and recently in 1990, when energy crises rose. But under normal circumstances, the public remains disinterested in energy issues. And the simple reality is that the public will prefer to keep energy prices low rather than raising energy prices through taxes.

When political leaders are gung-ho and say, pay higher taxes to induce energy conservation, or pay higher taxes directly to make investments, or raise prices, or some other thing that involves sacrifice, the public is going to resist that, save in those rare periods in which it is worried about energy. The normal circumstance is for the public to resist these kinds of sacrifice, and the quick and successful results in the Gulf War suggest to the public that there's no reason to make those sacrifices. It seems unnecessary when you just demonstrated that no one is going to be able strategically to dominate the oil resources in the Middle East. And so the downside, as it were, of the enormous success of the Gulf War is that it undermined the budding interest in energy policy.

Q: In 1994, the Rocky Mountain Institute estimated that the oil and nuclear energy received a $24 billion worth of taxpayer subsidies. Do you see a role in removing subsidies?

Schlesinger: Well, it's never been clear whether the competing forms of energy are interested in removing those incentives or reallocating those incentives so that they get a bigger share. For example, the solar industry for awhile wanted to have a substantial government program reducing the cost of solar energy. Your question goes beyond the issue of reallocation – just get rid of them all. That may run up against another government objective, which is to limit dependency on foreign sources of supplies. Removing the subsidies would indeed raise energy prices, and would result in a greater degree of energy conservation, but it would impose a greater burden on our domestic industries.

Now, there is a lot of talk about leaving the future of domestic energy industries up to the marketplace. But that tends to be just talk, and the reason is that the United States has a comparative disadvantage in the production of oil. It costs you $2 to produce a barrel of oil in the Middle East; it costs you $13 to produce it in the United States, or $18. Strict application of market forces, U.S. domestic oil production would go down to about 3 million barrels per day. If one believes in the marketplace, that's all right because it would indeed lower the cost to the consumer. The effect of getting rid of domestic industry though the end of subsidization will be to lower the prices which the consumer pays for petroleum, but that is not desirable on national security grounds. So we have a conflict there.

Q: As a homeowner, I recently surveyed local suppliers of furnaces, to replace the one in my home. We found that it's almost impossible to buy a high-efficiency one from these small, local contractors. They want to push the 80% versus the 90%+ current, and I find that rather interesting. It's a very easy way to save a little money, but they're pushing the fact that they have higher maintenance costs with higher efficiency models, etc. And you have to fight them to get a high-efficiency furnace. I mean fight.

Schlesinger: Well, that is an interesting reflection of some of the inherent problems in the marketplace. It is assumed by economists who play around with the theory of the market that consumers are knowledgeable and they have demands. That is not the case. One of the reasons that U.S. industry has been relatively successful in energy conservation is that industry indeed is a demanding consumer, and industry is quite knowledgeable about the ultimate costs and benefits of investing in more fuel-efficient equipment or appliances. The household consumer is not, and the consumer is swept by advertising and considerations of cost, as you just mentioned, or a host of other things, and the consumer is not necessarily wrong. This has led to continuous argument about whether or not it is appropriate to set minimum standards for appliances or furnaces, and there has been great reluctance to do that. Management desires to leave it in the marketplace for practical and ideological reasons. There has also been the desire to leave it alone simply because of the impact on those business firms that sell less efficient devices. As long as one leaves things to the consumer, that consumer who is not knowledgeable, is not a demanding consumer, will be swayed by these kinds of considerations. I support establishing minimum efficiency standards for the appliances that consumers use in their homes.

FUNDING AND IMPLEMENTATION OF THE EXECUTIVE ORDER ON FEDERAL ENERGY MANAGEMENT

Robin Roy, Senior Analyst and Project Director, Energy and Materials Program, U.S. Congress Office of Technology Assessment.
Mike Synar, U.S. House of Representatives (D-Oklahoma).
John P. Millhone, Deputy Assistant Secretary, U.S. Department of Energy.
Michael D. Beugg, Energy and Science Division, White House Office of Management and Budget.

Moderator: The difficult question to be addressed by our panel today is, how do we generate the necessary federal funds to implement energy conservation? How do we fund executive work? There are many ideas, including increasing agency budgets, shared energy savings, utility rebates, a bonding authority by the Department of Energy and, just recently, the energy conservation bank legislation introduced by Senator Herb Kohl of Wisconsin. There are many good ideas here, and what we'd like to do is to generate some further discussion of how we move forward on this issue. Today's panelists are specialists. They are leaders in energy conservation policy. They will offer their insights on the next steps we need to take to promote energy conservation within the federal government.

Robin Roy: Fifteen years, that's what we have. Fifteen years of agency programs, executive orders and legislation, suggesting, imploring, and mandating the improved energy efficiency of the federal government. What has that long effort brought? For one thing, it's brought considerable savings. According to the Department of Energy, between 1975 and 1989, federal efficiency efforts saved about $7 billion. That's about 5% of the direct federal energy spending at this time. It's also about 2-3 times more than was invested in energy efficiency investments during that time.

Yet despite that long effort, a large opportunity remains on tap. The federal government is the nation's largest energy consumer, spending in 1989 about $9 billion on energy in its own facilities and operations, and about another $4 billion subsidizing the energy expenses of moving and households. Much of this energy is still inefficiently used and there are many excellent opportunities for cost-saving improvements. For example, it appears that virtually available cost-effective technologies could conserve on the order of 25% of the $3.5 billion spent on energy in federal buildings. No sacrifice to health, comfort, or productivity. This has not resulted in darker offices or cooler offices in winter and warm offices in summer. These aren't exotic technologies and they may not be here next year, or in ten years. Rather, these are technologies which are commercially available now and in limited use now. Technology such as sensors that turn off lights when they're not in use, high-efficiency lamps and ballasts and fixtures for lighting, and improved operation and maintenance of heating, ventilating, and air conditioning equipment, often with the aid of energy management control systems.

It's important to note that the best federal practices are very good, and the more thoroughly implemented, are open to very high levels of efficiency and cost savings to the government. These practices include basing capital investments on life-cycle cost analyses, performing maintenance and operations efficiently, following the energy efficient building standards for new federal construction, establishing incentives for facilities and for individuals to perform efficient operations in maintenance, and participating in efficiency programs offered by local electric and gas utilities. A basic challenge is to make these best practices the standard practice. The natural question remains why, after 15 years, more than 15 years of agency programs, executive order and legislation, why is it that there's still so much untapped potential? Why are the best practices still the exception rather than the rule? So much the exception, and why are the best practices so different from standard practice?

Overall, on an abstract level, energy efficiency receives a relatively low priority. That's not surprising, as Dr. Schlesinger said, given that energy efficiency is non-fundamental to most or almost any agency's mission. It's also important to note that total energy spending is a very small fraction of any agency's budget, 1-2% or less. So many excellent cost-saving opportunities in energy efficiency don't receive the priority. It's understandable that they receive a low priority, but it's also worth asking, if we can't save federal funds, federal spending, here, where the opportunities are great and the technologies are relatively low-risk and well demonstrated, where can we?

The executive order and continued congressional interest and other activities helped raise priorities, but are not a mild solution of any sort. One place that low priority is seen is in a critical shortage of capital to make investments, even rapid payback investments. Many investments have paybacks of 100%. They pay themselves back in the first year. Particularly when they are made in conjunction with utility rebate programs. And we find these are going untapped. Capital budgets earmarked specifically for these types of energy efficiency projects in federally owned facilities dropped from a high of about $300 million per year in 1981 to $150 million in 1990. That's declined about 90%, and, in real terms, after inflation.

Some of this has begun to reverse with the General Services Administration (GSA) and the Department of Defense (DOD) alone increasing their efficiency investments from about $7 million for the two agencies in fiscal '89 to about $40 million for fiscal '91. Still, even with increases which are planned by DOD and GSA for the next several years, the investment level is far lower than it was in the early 1980s. You know, we've made profitable opportunities on-tap. How much could you probably invest? Easily several billion dollars could be properly invested in cost-effective energy efficiency measures. And what we're using today is currently old technology. Each year these investments are deferred, opportunities are lost that cost the federal government hundreds of millions of dollars.

A second place where there are low priorities, and perhaps more importantly, is in personnel. Adequate funding alone is not enough to get a high level of efficiency and cost savings for the government. It's at least important to have trained, knowledgeable staff at facilities to select programs, to work with the vendors, to work with the utilities that are offering efficiency programs and incentives, identify measures and to see that those measures are implemented properly. We want to avoid unpleasant effects as were found in the early '80s, late '70s, where some buildings were made very cold, lights were changed and offices became quite dark in many places. Having an adequate staff to identify those programs or projects and see them implemented properly is essential.

The types of problems facing the federal government are substantial, but they're not overwhelming. We found the best practices out there are addressing things such as giving the federal facility personnel incentives, giving the agencies overall some types of incentives. We find that life-cycle cost methods are not widely implemented. Better implementation of existing programs, finding ways to improve the programs, making the best practices we find today standard practice, can go a long way.

Several bills are pending in Congress right now. The Executive Orders signed last April are very helpful. These efforts will provide valuable support and guidance to agencies in pursuing that greater efficiency. However, based on past experience, on long history, it's best to think of these latest steps as an ongoing long endeavor, rather than a final step guaranteeing future performance.

Mike Synar: I am delighted to participate in this forum on energy conservation. I think those who follow my career know this is an issue that is very near and dear to my heart. I refer to the Executive Order on federal energy efficiency. As Doug Decker said, my chairmanship on the energy environment and natural resources subcommittee in government operations, along with my membership on Chairman Philip Sharp's energy and power subcommittee, has really put me in a dual role of not only doing proper oversight in this issue but, more importantly, implementing the legislation that can work.

We held a hearing in July of 1990 on energy efficiency in federal buildings and facilities. The hearing really was an effort by those of us in Congress who examined the federal government's attempts to reduce its own energy use and comply with the requirements of the Federal Energy Management Improvement Act of 1988. That hearing was a real eye-opener, because what we found is that the nation's number one, single largest energy user is our federal government. It's spending about $8 billion on energy, and about $4 billion of that was just in the energy cost of the government buildings and facilities. We heard testimony from fine people like Doug Decker at Johnson Controls, and we highlighted some things that I think the American public weren't aware of. We showed that the federal government has made an abysmal effort to reduce energy consumption. Instead of being a leader in stimulating the market for energy-efficient goods and services, the federal government has been almost the opposite. I don't need to tell you that the two major users of energy are DOD and DOE, they really need the most attention of all the agencies we examined. Because we let President Reagan's 1985 executive order lapse, what we've seen is that the agencies got very little incentives whatsoever to pursue implementing energy-efficient technology. We stepped it up to the plate to try and deal with this with the Federal Energy Management Improvement Act of 1988. As you are all aware, it required an additional 10% reduction in energy use by 1995.

Unfortunately, what we found in that hearing was that the act was not being implemented to the fullest extent possible. In fact, the federal energy use had actually increased from 1985 to 1989. Our subcommittee has identified a variety of disincentives that stand in the way of achieving this energy efficiency. It includes low federal funding in the energy efficiency initiatives and the difficulties of federal procurement, it's just really a general lack of commitment to push energy efficiency at the very top ranks of the executive branch. We found that there were very few federal installations that participated in energy savings programs. Even where such programs were substantially financed by someone else, we couldn't find much action.

Consequently, we pressed hard in our hearing, Phil Sharp and I both did, for the President to issue the executive order requiring greater energy savings. This new order, signed on April 17, 1991, now requires a 20% reduction by the year 2000. Why does this make sense? Well, from a public policy view, it's clear that this efficiency will lessen the disruptions of our energy supply. It will extend the life of our oil and gas reserves. It will slash emissions of harmful sulfur oxides and nitrogen oxide. It will reduce our output of gasses which are responsible for our global warming problems. And energy efficiency will really pay for itself, and within literally months and years.

Well, let me bring you up to date, because Ed Markey, if he had been here today, would have told you about what we're trying to do to the national energy strategy which is working its way through the House right now, and should be completed within the next couple of weeks. Because what we're trying to do with the national energy strategy is to recognize that this is an important part of the solution. The federal facilities title which will be in the national energy strategy contains extensive provisions for increased energy efficiency. It will require federal agencies to implement all energy efficiency improvements that have a ten-year or less payback. It sets up the energy efficiency fund to help make those improvements. It rewards facility managers who have done an outstanding job of saving energy, and it authorizes federal agencies to participate in utility-sponsored efficiency programs. This is a giant step forward from where we have come in the past.

The national energy strategy conservation title, which is something Mr. Markey and I have been working on, includes a lot of things. For example: Federal Energy Savings Incentives Act will correct a flaw that we uncovered in that hearing, and provides for incentives that we agreed were necessary to achieve significant energy savings. It allows an agency to keep 1/3 of the savings generated by installation of energy-saving equipment, and the money can be used for what we call agency moral welfare programs, such as recreation, child care, or continuing education. Another 1/3 would be reinvested in energy conservation methods to generate further savings in energy and tax dollars. And the final 1/3 would go to the Treasury for deficit reduction.

Now this is a system that has already been tried. This is not something that we need to know how the results are going to happen. Because it's being tried right now in the Defense Department, and the federal government's chief energy-using branch is showing it can be a big success.

The national energy strategy will also deal with other problems that we uncovered in our hearing. Most important are obstacles to federal participation in utility-shared savings plans, under which utilities pay the initial cost of installing energy-saving equipment, in return for receiving a portion of the agency's fuel savings. Our bill is going to deal with that. We would streamline the procedures for agencies, and what we would ask is that they take part in the shared savings plan. We would make these programs more attractive, both for the federal government and our utility systems.

But the issue at hand is really the issue you started today in discussion of this, and that is, where are we going to get the money to pay for it? Now I'm going to listen with great interest when OMB comes up here and talks in a few seconds, because what we're going to have to have, literally, from the highest levels of federal government, is some kind of national commitment. From the high of federal spending in 1981, funds for efficiency have literally been cut 85% by 1985. Now while they've increased somewhat recently, there has been a change in the direction. I think all of us who've been involved with this realize we have a long way to go.

Overall, I think you can generally assume that the energy strategy which will finally become law will have a major component of conservation. And those of you who are all for this idea need to get prepared for two things I think will happen in the 1990s. First, we will make sure that the American public knows that the cheapest energy that we can have in this country is through conservation. And secondly, the federal government need not follow the lead in this area; and finally that Americans will become aware of the fact that through conservation we can depend upon ourselves more and less on foreign countries.

John Millhone: I'm going to first provide some basic information on what the record has been as far as energy use by the federal government is concerned, and talk about some of the things that already are being done. Then I will talk about the need for some bank or some source of capital funds for the kinds of investment in energy efficiency we need in the federal government.

The cost of energy used by the federal government last year was nearly $10 billion. This is reported by the federal government in two broad categories: one is in buildings and related process activities, and one is in general operation, which includes aviation and other operation facts. The largest energy use is in this general operations category. The government uses nearly a full quad in general operations. More than 90% of that is used by the Department of Defense. I imagine that with Desert Storm, that increase is going to be up sizably during 1991. In the buildings category that is the primary focus of the bank proposal, the amount of energy used was nearly about half a quad. That's the site energy. Exactly 447 trillion BTUs. This cost was about $4 billion, which was 5.5% less than the amount used in buildings for 1985.

In the buildings category, the National Energy Conservation Policy Act requires all agencies to take actions that are necessary to reduce their energy use per gross square foot by 10% from 1985 to 1995. The progress so far has been a reduction of 5.8% per gross square foot in 1990. So we're making pretty good progress toward this goal.

This is energy awareness week. We have identified the 43 best performers last year as far as energy efficiency in federal buildings is concerned. Those 43 winners have produced $243 million in savings. That's a result of just the people who got recognition last week for what they've done. So there's been significant progress already made in terms of energy efficiency in the federal sector.

We have taken some steps within the last two years to make further significant gains. The National Energy Strategy (NES), in February called for a much more aggressive federal government effort to exercise leadership by increasing energy efficiency in building design, in operations, and in procurement practices. The NES goals, added to the executive order issued in April, call for an acceleration of what already had been the aggressive approach that Congress had indicated the federal government should take. It calls for a 20% reduction in energy use from 1985 to 2000. Before the requirement was only for energy use per gross square foot in buildings. Now this has been extended to industrial activities of the federal government. We are looking at different industrial processes, and how production per unit can be improved in terms of energy efficiency by 20% during this 15-year period.

In addition, there were calls for accelerated work on alternatives. You will hear about bills addressing an effort to lead in the procurement of energy efficient products. The Department of Energy has done some things we think are important in this area. We've increased the status of the Federal Energy Management Program (FEMP).[1] We've increased the budget for the activity from $1.2 million in 1990 to a request for $4.5 million in 1993.

---

[1] The DOE Energy Bank study is to be released in March1997. This concept is being re-introduced into the 105th Congress in 1997.

Congress appropriated $4 million,[1] we got twice the support from Congress than we expected for expansion of this activity, and there has been a good deal of work in other departments as well, in increasing the funding for energy efficiency investments by DOD and GSA. Last week, GSA presented a report about energy savings goals. It published in one document all the information they could find about energy-efficient products. The report has gone to all federal acquisition people so that they know where they can get the more efficient products.

So a good deal of work is being done now to improve on the record that's pretty good to start out with. But there are additional things that can be done and should be done. One of the things that needs to be addressed is how do you obtain the funds necessary to make capital investments that are required to accomplish the energy savings. We recognize that when an agency or department has to make a choice between investing in something that advances its mission, or something that improves energy efficiency, it will opt for the investment in advancing its mission. And it should. That's the primary objective of that particular agency. So there needs to be some way of providing the funds necessary to make the additional investments in federal energy efficiency.

There are three ways that we get funds: there are funds that come from the appropriation requests; there are funds which come from participation in utility-funded programs; and there are funds that come from shared energy savings programs. These can provide significant funds. We need to look more carefully than we have in the past at other sources, or at least look at what the funding requirements are so that we can accomplish the kind of cost-effective energy savings that are still out there.

From the Department of Energy, we see this proposed agency bank legislation as a good contribution to the resolution of this question. But we have some questions. For example, is it possible to pass this legislation and guarantee that the appropriations committee will appropriate the funds necessary in addition to the normal utility costs, to fund this conservation bank? The bank provides some arrangement where there will be payment by the department, but it's usually going to pay interest as well as the principal amount. Is it reasonable to have one department in effect, take the interest of another department in order to preserve this bank? Is the $200 million that's set forth a reasonable amount? Is that the amount that's needed, is it too much or is it too low? We are quite willing to work with Congress and with Johnson Controls to find the best approach in dealing with these kinds of questions.[1] We're very concerned that when you come up with something like this, that you don't turn off the other sources of funds. If it would be possible for people to say well, with this fund that is going to be created, that's where the federal government will meet all the capital requirements that it needs, and so therefore we can use some of the capital funds that we otherwise might send in this way in some other direction. There's a real temptation to try and save any place you can. So we're very concerned that the manner in which this is addressed carefully sets forth a program that will work.

In our budget for the current year, we at DOE requested funds to undertake a study to identify what the needs would be for capital equipment, capital investment, in energy efficiency. We received this funding from Congress. We will be undertaking that study during the year. We want to work with OMB, with congress, with GSA, with DOD, with the principal stakeholders in this area, to make sure that we look at this in a way that comes up with a realistic program that will work.

Michael Beugg: I'd like to start off by thanking Johnson Controls for their support throughout the entire road to getting the executive order issued. They played a very valuable role in that process, and we at OMB and elsewhere throughout the government certainly appreciate that. Now that being said, I had to wonder when I saw the composition of the panel, and likewise seeing the audience, whether this isn't some kind of a set-up. Here I am from OMB. We're the only organization in the world, I think, where you can earn your pay just by saying "No" to everything, all the time. As most of you know in the private sector and elsewhere in the government, it takes more than saying 'no' to succeed. You have to come up with creative solutions to problems. But with 5 million federal employees who like to say 'yes' to most things, we can afford to have about 500 of us who say 'no.' That's why there's an OMB.

All of us here on the panel today, whether from the legislative branch or from the executive branch, as well as those from the government and industry, share the same goals. We want to protect the environment. But today's session really isn't about discussing those goals, I think there's a lot of congruence on the goals. What we're really interested in talking about is what are the best strategies, and what are the best funding mechanisms to get us where we're going. Our two major choices are administrative action coupled with agency budgeting and the appropriations process; or broader legislative solutions. This is, after all, a management problem, and I think President Bush when signing the executive order said that, "After all, this kind of thing is just a good common-sense business practice." That suggests that at least those of us in the executive branch might want to advocate the first solution.

I think that, as John Millhone has pointed out, the system is working considerably better than many of those who speak on the issue of energy efficiency might indicate, particularly because of the confusion resulting from site-based reporting versus source-based reporting. We tend to hear a lot of statements in the press like, "The government is failing in energy conservation, or energy consumption by the government is rising"; whereas, the fact is if you measure at the site, which, after all, is what the energy manager is responsible for, there has been a 25% decline in energy use by the government between 1975 and 1985, and another 6% decline from 1985 to 1989. Now, that is relatively good. Certainly we can do better. But saying that we aren't doing a good job at all, or that the federal government's efforts to date are disgraceful or vastly inferior to the private sector, seems to me to be a bit off the mark.

You've heard about how federal energy efficiency spending has dipped, and it has dipped from the mid-70s and late 70s to the early 90's, but now it has gone back up again. My calculations are that in excess of $100 million per year will be spent by the big four energy consuming agencies, and that's not including the Postal Service: those are the Department of Defense, Veterans Administration, the Energy Department, and GSA. So if you look, not necessarily at the line items in the budget book, but into the operations and maintenance budgets of those agencies, and you ask them how much they are planning to spend on energy conservation, you'll find that from fiscal year 1993 essentially through the end of the decade, numbers in excess of $100 million are planned right now. That's possibly enough money to achieve the energy savin goals of the executive order. It depends on the average payback of the investments involved; it depends on the amount of utility cost-sharing that occurs. Later in the decade we may want to revise those numbers upward, but these are significant numbers.

Now, with regard to the fund for energy efficiency, I think there are a lot of good things to say about it. I like the idea of more business-like management practices in the federal government. But at the same time there are certain problems with this particular approach, which have come up in the formulation of the national energy strategy, and which will doubtless come up again. The good points of the approach are: it establishes excellent project selection criteria for evaluating between a multitude of proposed projects, and I think anybody can be happy with that. It's also a very good way to gain the sort of portfolio management benefits that those of us who study stock portfolios are familiar with. The diversification of multiple project selection eliminates the risks associated with unpredictability, or diminishes it at any rate, so that's good too.

The problem is with the inter-agency nature of this fund. If you take a look at federal energy management abilities and how it's distributed, about 74% of that is in the Department of Defense. The question then raised by this federal energy efficiency fund is – why, if one of the agencies accounts for 75% of energy spending, would you want to "tax" that agency to create the fund, and then force them to make project applications to people who manage the fund to get that money back again? I'm not sure how that really increases the efficiency of availability of funds from the government. In other words, 75% of the spending, or $3 billion worth of buildings and energy expenditures, seems like an adequate base over which to develop effective project selection criteria, and to diversify the risks of those projects selected within an individual agency.

So while the federal energy efficiency fund idea has a lot of promise, it seems like something that would apply best on a much smaller scale to coordinating the project selection activities of a small agency, or a group of smaller agencies. I don't see that the Department of Defense, in this era of declining budgets, would be particularly pleased to have, something on the order of 6%-7% of its energy bill taxed, going out to create the fund. In this case, that's about $150-200 million. This will be deficit-increasing eventually. There's no two ways around it. Don't let the fund fool you, the money has to come from somewhere. It's going to increase the deficit in the near term; it's going to increase the short-fund account, and it's probably going to squeeze out the DOD energy management activities right now. Their planned annual funding is $50 million, and maybe that's not enough. Certainly if you take $200 million away from DOD's energy managers, though, they're not going to do a whole lot more energy conservation spending on their own. They'll have to wait and see if their projects are selected by the fund managers. So I'm concerned about that aspect of it.

The other thing that I'm a little bit concerned about is that there's no "sunset" provision of the fund. As we all know, government programs. And government structures, once enacted, tend to go on long after their purpose is served. We'll probably have to spend money on energy efficiency for a good bit of time. But the question is whether funding at the level specified by the bill, a minimum of $200 million a year, is always going to be appropriate relative to other pressing needs that the government has for health care, transportation, etc. Energy saving, even when cost-effective, while a worthwhile goal, still has to be measured against other equally cost-effective programs. So, once set in stone for eternity, a level of funding for federal energy efficiency doesn't necessarily, every year, make the best choice for the trade-offs that our tight budget environment is going to present us with.

So those are my basic points. There are some other interesting issues raised by the bill; like "securitization" of energy investments, and about the role of GSA in managing properties of agencies that rent from them. For instance, should the agencies take primary responsibility for those buildings, or should GSA with its greater expertise in energy management? But those are not the major concerns. This shouldn't obscure the idea that funding for federal energy efficiency is necessary, and we at OMB support that. And lastly, there are some benefits to having a bank-like structure like the proposed federal energy efficiency fund. The question is, just how broadly it should apply to which federal agencies. And with that, I'll conclude.

James R. Schlesinger, former Secretary of Energy and Defense, delivered a sobering address in 1991 that focused on the interconnections between energy conservation and our position in the world.

Rep. Mike Synar (D-Oklahoma) discusses the Executive Order on Federal Energy Management at the 1991 forum.

Robin Roy of the U.S. Congress Office of Technology Assessment discusses a 1991 report indicating the government can save $1 billion per year through energy efficiency in buildings.

Third Annual

Energy Conservation Forum

1992

Third Annual Forum – 1992

Beginning with the September 1992 conference, the annual forum expanded in several dimensions. The United States Energy Association (USEA) joined Johnson Controls, Inc. as a sponsor of the event. The USEA, an association of nearly 200 public and private energy-related organizations, is the U.S. member of the World Energy Council (WEC), an international non-governmental organization founded in 1924 to study, analyze, and discuss global energy issues.

USEA's co-sponsorship brought with it the first Foreign Study Tour, a group of energy leaders from developing countries who learned a great deal from attendees. The third annual forum also inaugurated a series of energy efficiency awards, presented to individuals from the public and private sectors in recognition of their outstanding work in the field of energy conservation and efficiency.

The 1992 forum included a presentation of the USEA report recommending a five-point energy efficiency strategy for the U.S., *Getting Down to Business*. In keeping with the missions of USEA and WEC, this forum covered international as well as domestic developments. One panel of American experts reviewed domestic issues, a panel of international authorities informed the audience about energy issues in their developing countries, and another panel of American speakers discussed U.S. perspectives on, and participation in, the global energy scene.

We reproduce here the remarks of our speakers from India and Mexico. Sadly, R.D. Aga passed away in early 1996, but the chairperson of the Thermax, Limited board of directors felt strongly that his remarks should be included in this book.

We also present the morning and afternoon keynote addresses by Michael Deland of the President's Council on Environmental Quality; and Linda Stuntz, Deputy Secretary of the U.S. Department of Energy, which focus on different aspects of the energy policies and accomplishments of the Bush Administration.

ENERGY EFFICIENCY: THE CORNERSTONE OF ECONOMIC GROWTH AND ENVIRONMENTAL PROTECTION

Michael Deland, Chairman of the White House Council on Environmental Quality (1989 to 1993).

I've been asked to help set the tone for the day and I welcome the opportunity to do so, since I'm convinced that energy efficiency is a cornerstone of our energy future, both domestically and internationally. Years of talking about energy efficiency have generated an irresistible momentum for action, and indeed there never has been as much activity devoted to energy efficiency as there is today. The United States is a leader in this regard, as I'm sure you know. Over the last 20 years we have lowered energy use per unit of economic output by approximately 30%. Energy-related air pollution has fallen, while our gross national produce has risen over 50%. Not a bad record. But today we should talk about and focus on the future. We already know that a growing economy and cleaner environment are inextricably intertwined and it's really that, that is the cornerstone underpinning of George Bush's philosophical approach to environmental protection and to energy efficiency. He has reiterated that on numerous occasions, in each of his annual messages to the United States Congress. We already know that energy efficiency is a part of what makes those goals compatible. We also know that energy efficiency is not about, as your film suggested, freezing in the dark.

Let's go beyond the debate of the last 10 or 20 years and ask what factors will drive energy efficiency for the next 20 years. I see at least two, and I'll focus on the two that I suggest are of the most import. The first of these is economic renewal. As the President explained in his Detroit economics speech earlier this month, the U.S. economy is in a period of difficult transition following the end of the Cold War. We face a huge national debt, we need investments in plant and equipment. We've got to win the tough economic competition of global markets. All of these features of the current transition rule out inefficient ways of doing business. In this economic climate we simply cannot afford waste. We must increase our productivity, and that means dramatic efficiency gains in every aspect of production and of consumption. The second factor that will drive energy efficiency in the coming decade is concern about environmental impacts of energy, and particularly global climate change. Like most of you, I'm a committed environmentalist and I'm concerned about loading the atmosphere with greenhouse gasses. While there are important questions still unanswered, such as the rate, magnitude, and geographical distribution of climate change, I do believe, as does the President, Admiral Watkins, Bill Reilly, and all of us on the Bush team, that there is enough reason for concern to warrant prudent actions now. The debate of the 1990s is not about action versus inaction, it's about what kind of action makes sense. And the successful negotiation of the U.N. Framework Convention on Climate Change signed by more than 150 nations in Rio last June, drives the point home.

Efficiency is at the top of the international agenda and that is not going to change. Earlier this month, the administration submitted the Climate treaty to the Senate and called for immediate ratification. Last week, in fact, I personally wrote to the Senate leadership on the President's behalf, stressing our commitment to this goal. And if Congress moves, the U.S. could become the first industrialized nation in the world to ratify the treaty. We expect the Senate Foreign Relations Committee will report out the legislation on Thursday of this week. Even now, the United States is one of only two governments in the world to have put on the table a specific action plan to deal with climate change. It contains strong energy efficiency measures and other programs, which collectively will reduce net greenhouse gas emissions by as much as 11% below projected levels in the year 2000. That's a strong and a realistic commitment. The administration worked long and hard to establish a Framework Convention that is ambitious but flexible, not prescriptive. The President, along with many of you, fought for that kind of a convention because it's the most effective and efficient way to protect the environment, while enabling all nations to raise their standard of living. A much more sensible approach than the CO_2 only avenue that many espoused.

So how do we get there? How do we create an energy future which reduces the impact of energy-related emissions, controls the threat of climate change, and fuels our economic renewal? Start with a national energy strategy. Deputy Secretary of Energy Linda Stuntz will expand upon this at noon today, but the President's National Energy Strategy (NES) will tap efficiency as the single greatest source of new energy, roughly equal to or greater than the amounts that will be generated from increased production. The President's strategy includes elements such as increased use of integrated resource planning, provision for performance based contracting, higher efficiency standard for buildings, appliances, equipment, and under the NES, the administration and Congress have already nearly doubled funding since 1989 for conservation research and development, and for clean energy technologies like solar, wind, safe nuclear, and advanced transportation technologies. The President signed an executive order requiring a 20% reduction in energy use in federal buildings and vehicles by the year 2000, saving taxpayers an estimated $800 million a year. To date, more than 90 energy strategy initiatives have been launched through administrative action, and we are doing everything we can to help Congress complete its energy legislation before it adjourns.

The National Energy Strategy is not the only program underway which will spur energy efficiency. The acid rain portion of the new Clean Air Act gives utilities the flexibility, and indeed, the incentive to reduce SO_2 emissions in a cost-effective manner. We expect that as utilities try to meet demand and stay within the cap on SO_2 emissions, they will adopt aggressive energy efficiency programs. The Clean Air Act also provides bonus allowances to utilities which achieve SO_2 reductions through energy conservation. We have launched several voluntary programs to supplement government action. More than 600 participants, for example, have joined the Environmental Protection Agency's (EPA) Green Lights program, making a voluntary commitment to install efficient lighting in 2.5 billion square feet of office space. That's more space than exists in New York, Los Angeles, Chicago, Houston, Dallas, San Francisco, and Washington combined, so not an insignificant program. In another voluntary program, EPA and a consortium of utilities have established what they call the Golden

Carrot. Not too long ago we were dealing only with sticks, but now the Golden Carrot from EPA. A $30 million prize for the first company to market a dramatically more efficient refrigerator, using substances that don't harm the ozone layer. By the year 2000 such refrigerators could reduce electricity consumption by 3 billion kilowatt-hours per year, saving consumers some $240 million on their electric bills.

Then there's another project, one in which I'm personally involved, the President's Commission on Environmental Quality or PCEQ. PCEQ focuses on results, not rhetoric, and who better to oversee our energy efficiency project then Dick Clark, the chairman of Pacific Gas and Electric. And I might parenthetically describe briefly how the commission was chosen. We selected first environmental group leaders, such as Fred Krupp, the head of the Environmental Defense Fund, Pat Noonan of the Conservation Fund, Katherine Fuller of World Wildlife Fund, and then reached out to corporate leaders who had demonstrated individually, and through their corporations, and through their trade associations, that they, indeed, were pioneers, so Jake Jacobsen of 3M, the father of the "Pollution Prevention Pays Program" is a member, and Dick Clark is a member because PG&E over the course of the last decade has saved 1,300 megawatts of power through its energy efficiency program, thereby saving its consumers $3.5 billion. In an effort to widen U.S. industry's involvement in energy efficiency, the Council is seeking companies and organizations that will invest in programs that show measurable results in one or more of the following ways: by investing in cost-effective efficiency improvements, by adopting a charitable contributions policy that assists grantees with energy efficiency so that more funds will be available for charity rather than energy bills, by encouraging employee involvement in energy efficiency, by reforming commercial leasing practices to create incentives for energy efficiency, and by encouraging the adoption of integrated resource planning by states and utilities nationwide. We now have some 200 different entities involved in the work of the Commission. The success of all these voluntary programs: Green Lights, PCEQ, and others, indicates the general magnitude of cost-effective efficiency improvements that are achievable in the near future.

I've just listed more than a dozen domestic initiatives to revolutionize our energy economy, but as I mentioned earlier one of the driving factors of the next decade will be concerns about the global economy and environment, particularly as developing nations work to eradicate poverty and expand their economies. If we do not help the developing world improve its energy efficiency, its energy sector growth will in a blink of an eye outstrip the available development assistance and commercial investment capital. Studies project capital needs on the order of $145 billion per year early in the next century. As on the domestic scene, however, we are starting to see some good news. Multilateral development banks are finally beginning to reform their energy lending. This summer the Unites States led other donor nations in urging the World Bank to promote the use of integrated resource planning in all its energy lending. Later this week the U.S. Agency for International Development, the U.S. executive director at the bank, and the executive director representing much of Central and South America, will cosponsor a seminar on energy efficiency and conservation experiences in the Central American region. On a bilateral basis the United States also devotes a tremendous amount of financial and technical resources to energy efficiency.

These are important government initiatives. However, government's primary role is to foster conditions that encourage private sector investment in advanced technologies. So in tandem with our growing economic ties to the nations of Latin America, the Pacific Rim, the former Soviet bloc, the United States has established new programs in each of these regions to cooperate on energy efficiency, renewable energy, and other environmental endeavors. Today I am pleased to announce a major new initiative to promote energy efficiency worldwide. Through the U.S. Agency for International Development, the United States will provide financial support for the Atlantic Council to establish a World Council for Energy Efficiency. The council will be a nonprofit public-private partnership to disseminate energy-efficient technologies to the developing world. A convening meeting will be held this fall and we expect this international clearinghouse to be based here in Washington.

We're entering a period which I expect will bring growth and dynamism to your industry. We're entering a period in which we aim to have a growing economy and an ever cleaner, safer environment. Those goals don't come about by just saying they should, or by imposing new taxes and regulations on a global economy that's already struggling. Those goals will come about through performance-based, market-oriented government policies and through active, forward-looking, innovative private sector leadership. Our energy efficiency strategy mirrors the one that you have proposed, particularly in mobilizing public and private leadership, making energy efficiency an engine for growth and spurring international technology cooperation. But we have only just begun. Our challenge, working together, is to consolidate the gains of the past few years, make them integral to the economy of the 1990s, and design the next round of innovation that will expand our options into the 21st century. There is much at stake. Millions of jobs in this country, around the world, our economic renewal and that of the developing nations, and the quality of the environment that we leave to our children. By working together we can indeed, lead the world in energy efficiency, and hence, meet those common challenges.

THE NATIONAL ENERGY STRATEGY OF THE BUSH ADMINISTRATION

Linda Stuntz, Deputy Secretary, U.S. Department of Energy (1992-1993).

I am pleased to be here and accept your gracious invitation to attend this important conference and celebrate the renewed interest around the world in energy conservation, and promoting sound policies that help us conserve energy and use energy more efficiently. I especially want to talk about the significant accomplishments of the Bush Administration in improving the energy efficiency of America. In one sense, I view this as a matter of setting the record straight. If you look at some media, some interest groups, and certain politicians, you would think that the President and his energy strategy have nothing whatsoever to do with conservation. Let me say clearly: this President is strongly committed to increasing energy efficiency, and I believe his record proves this. The fact that his energy strategy is a balanced one, and includes both incentives to increase domestic energy production, and incentives to increase conservation, does not in any way detract from his efforts to develop aggressively our nation's conservation resources. The simple truth, as most of you recognize, is that we need both responsible domestic energy production and all the conservation that our wits and technology can bring us, if we are to meet our nation's and our world's need for secure, clean, and affordable energy in the 21st century. Those who would have us rely exclusively on one path or the other, or try to pit one side of the energy equation against the other, are really missing the boat, and they would do this nation a grave disservice.

First, let me dispense with a myth that is particularly annoying to me: the myth that Americans are the world's biggest energy wasters. Over the past two decades, energy use per household has dropped more than 25%, energy use per square foot of commercial floor space has dropped more than 20%, and industrial energy use per unit of output has dropped a full 50%. From a different perspective, between the years 1970 and 1990, the U.S. reduced its overall energy intensity, defined as energy consumed per dollar of gross domestic output in constant dollars, by 30%. Comparative statistics for Japan, France, and Germany over the same period of time, were 31%, 24%, and 22% respectively. Thus, in the last 20 years, U.S. energy consumers, with the help of companies like Johnson Controls and many others who provided the technology and the services, have helped save nearly one trillion dollars by investing in energy conservation and increased energy efficiency. It is true, of course, that energy use per capita in the U.S. is greater than in Europe and Japan; but our population density is far lower, which means that products have to be hauled farther, and people have to travel farther to do business. European and Japanese homes are also smaller, and tend to make less frequent use of central heating and air conditioning. Nor do they usually contain refrigerators and freezers in the numbers and sizes that we find them in American homes.

Can we become more efficient? Of course we can, and all of this is not to say that we should not attempt to do that. The key is to build upon our strengths: technology, and a free market economy. That is exactly what President Bush's National Energy Strategy (NES), issued in February of 1991, seeks to do. One part of the NES, that we all tend to focus on, is the legislative recommendations it makes regarding energy efficiency. Many of those are now pending on Capitol Hill, as the Congressional conferees struggle to find final agreement this week on sound, comprehensive energy legislation that the President can sign.

Before I talk about that, I want to turn to the other part of the NES that tends to get overlooked, because it's the things we do day in and day out in the bureaucracy. That is the things that do not require legislation, and that can be done through affirmative, executive branch action and leadership. One obvious example is the President's Executive Order on Energy Efficiency and Alternative fuels. Issued in April 1991, this order directs the federal government, which after all is the single largest consumer of energy in this country, to reduce its use of energy by 20% by the year 2000. If we meet that goal, we think we would save 100,000 barrels per day of oil equivalent, and nearly $800 million per year in energy costs. We're moving to implement that. The Social

Security Administration (SSA), for instance, has already completed energy saving measures that have produced over $600,000 in utility rebates. The General Services Administration(GSA) has established a $30 million fund to invest in energy efficiency measures.

DOE is doing its part under the Federal Energy Management Program, and has formed what we call a "test bed demonstration program" to demonstrate in federal facilities new energy-saving U.S. technologies. The program is administered by our Pacific Northwest laboratory, which recently signed its first cooperative research and development agreement with four organizations to test natural gas cooling systems. How long have we been trying to get those into the marketplace? Three private organizations are going to test the efficiency of gas-powered rooftop air-conditioning units developed by ThermoKing Corporation, which we think would yield significant return on investment over conventional systems. Our thought is, if we can actually get them demonstrated, help the architecture and engineering and construction community become comfortable with them, and demonstrate their performance, then we can attract the private sector. That's the way the federal sector can provide leadership.

Insofar as the DOE is concerned, let me reiterate that the President has increased DOE's budget for energy conservation R&D by 100%; he has more than doubled it since he took office in 1989. He has requested $366 million for 1993, and it appears the Congress will come close to approving what the President requested. In building technologies, the President's budget request for 1993 is $52.5 million, up 11% from last year's appropriation. $3.5 million of that increase is for appliance and lighting research and development. This is significant because appliances consume more than 40% of all the energy used in the residential sector, and lighting consumes 25% of the energy used in commercial buildings. Energy expenses of our nation's housing reach about $100 billion each year. As this group knows, the technology exists, or is being developed, to dramatically reduce that expense. The question is getting it used. Much of it is not being used right now, because the construction industry, the financial system, and homebuyers, all traditionally seek to reduce up-front costs of new homes, or improvements on existing homes. To help overcome this up-front cost barrier, DOE and the Department of Housing and Urban Development (HUD), the housing industries, and the financial community, are working together on a program of voluntary home mortgage incentives that would be linked directly to a credible home energy rating system. Under this program, buyers could apply for an energy efficient mortgage when purchasing an energy efficient home, or by planning intermediate energy improvements in an existing home. The idea is to encourage builders, buyers, and bankers alike to look at the total cost of the home, the total cost of owning an energy-efficient home, when figuring out what kind of mortgage a person could qualify for.

In a related effort, we have responded to the growing need for standardized conservation objectives in the building industry by publishing for comment, in the Federal Register, a proposed set of model residential energy conservation standards. These standards would be voluntary. They would not establish nationwide requirements per se, which we think would be mistaken and misguided in a country as diverse as ours, but they would encourage the generation of localized building standards, unique to the climate, cost factors, and construction practices of a particular area. The objective once again is to promote the design and construction of energy-efficient, cost-effective homes, apartments, and mobile homes. We're also collaborating with HUD to make federally assisted housing more comfortable and energy-efficient. For example, we have joined forces with HUD and Boston Edison to upgrade public housing in Chelsea, Massachusetts, to take advantage of energy efficiency technology. DOE and HUD will be providing the technical expertise and assistance to identify energy saving opportunities. Boston Edison is going to finance the improvements as part of its least-cost utility planning program. Our goal is to achieve a 25% savings, and if we're successful, over the next decade the energy cost savings potential of this initiative is estimated to be nearly $1.5 billion per year. So we're not only saving energy, we're helping to make public housing more affordable. That's a win-win if I ever heard it.

Yet another area where the NES has boosted conservation efforts is in promoting changes in utility regulation that encourage utilities to invest in conservation as a means of meeting their customers' energy needs, in addition to generation. In accordance with the NES, the Internal Revenue Service in 1991 issued a ruling clarifying that rate discounts provided by utilities to encourage customers to invest in energy-efficient equipment would be tax-free.

That's only part of the picture. Some recent analyses by Eric Hirst of Oak Ridge National Laboratory of utility demand-side management (DSM) programs, based on Energy Information Administration data published in May 1992, indicates that 30% of the nation's largest utilities, the 1200 or so with sales of more than 120 gigawatt-hours per year, ran demand-side management programs in 1990. These utilities spent $1.2 billion on their DSM programs that year. This is a big number. But the exciting thing for this group is that projections show DSM budgets growing to $2 billion per year by 2000. That's an average real growth of 5% a year, and I would suggest that even this aggressive growth is understated. For those of you who produce and sell equipment and services to save electricity, this represents a tremendous market, and a tremendous opportunity. A key emerging issue is the need for improved methodologies to measure the results of these programs, what's working and what's not, what kind of a payback are we getting for what kind of an investment? DOE is focusing on this question increasingly in our own Integrated Resource Planning, which the President has, again, substantially increased funding for during his administration.

In the area of transportation, this administration proposed, negotiated, and signed major new transportation legislation, increasing federal funding of mass transit, and giving state and local governments flexibility to fund other energy conserving transportation projects. The tax-free subsidies that employers may provide to employees were increased to $21 per month by administrative action from the Department of Treasury, the maximum we deemed possible at that time. In addition, the President, in his 1993 budget, proposed to increase that amount to $60 per month. We think in the transportation sector too, regulation is not the answer, but rather new technology and paving the way for its use. In FY 1992 we spent at DOE about $128 million for research and development on new engines and alternative fuels. That includes $39 million for new technologies such as ceramic engines, fuel cells, and gas turbines; $29 million for biofuels development; $43 million for electric vehicles; and $17 million for other technologies, including compressed natural gas vehicles. On the demand side, we have made substantial progress in supporting the purchase, and conducting the analysis, of alternative fuel vehicles, involving hundreds of cars, trucks, buses, and vans all across the country. Of the 3500 of these vehicles we've already purchased as part of the federal fleet alone, that is only the beginning. Thanks to the President's Executive Order on Federal Energy Management, we will buy at least 5000 more alternative fuel vehicles in FY 1993, and by FY 1998 fully 50% of all new federal vehicles will either be fuel-flexible, or dedicated alternative fuel vehicles, including some electric vehicles.

Finally, in the industrial sector, the administration is seeking to improve American manufacturing competitiveness through reduced energy requirements and waste minimization, both of which can lower overall production costs. One successful avenue for doing this that we're very excited about is the President's National Technology Initiative. This seeks to place the tremendous research and development capacities of DOE's thirty national laboratories, and really the entire federal government's 700 laboratories, more readily available for cooperative research projects with private enterprise. This is not just pie in the sky. Let me give you one concrete example. Right now DOE is jointly funding with the American steel industry a major program to develop a competitive process for making steel without the intermediate coking step. That's the energy intensive step, that's the step that creates the most environmental impact. DOE has invested over $45 million, and the American Iron and Steel Institute over $15 million, in a 120-ton per day pilot scale research project near Pittsburgh. We're planning to scale up this technology to a 1000-ton per day facility over the next four years.

In a different vein, our industrial audit program is geared toward maximizing use of today's most energy-efficient technologies. We now have 22 universities serving as energy analysis and diagnostic centers around the country, by the way also training engineering students at those universities in the skills and techniques necessary; they can then provide energy conservation and productivity audits to small- and medium-sized manufacturing plants who don't necessarily have the in-house capability to do this. A typical audit recommends quick payback measures that will usually yield savings of about $40,000 on annual energy costs.

Most often, about half of the recommendations are actually implemented by the firms, resulting in an average one-year payback, an annual savings, of about $20,000 for each company.

Another NES initiative in the industrial technology area was announced this past May, when we launched a ten-year, $100 million program to develop continuous fiber ceramic composites; i.e., materials that are stronger, lighter, more corrosion resistant, and better capable of performing at higher temperatures, than materials in use today. They will allow many industrial processes to be more energy efficient, and could save us up to 1.1 quads of energy, just in the uses that we've already identified. We've already awarded contracts under this program to ten industry-led research teams, which will involve private companies, universities, and national labs.

Finally in this area, let me tell you about NICE – the National Industrial Competitiveness through Efficiency, Energy, Environment, and Economics program. It's an exciting collaboration between DOE, the Environmental Protection Agency, and the private sector, actually working together. It doesn't happen that often, but I hope it's the wave of the future. This pilot project is based on cost-sharing partnerships with states and industry, and is currently operating in seven states: California, Illinois, Louisiana, New Jersey, New York, Ohio, and Texas. Though it's only been underway a brief time, this program has yielded successful results. For example, in a project with PPG Industries of Ohio, we have managed to reduce painting waste from 400,000 gallons to just 20,000 gallons per year. If any of you have ever tried to dispose of paint left over from household projects, you can imagine what that means to the waste disposal issue. We're very proud of this, and other waste minimization efforts, because they reduce the cost of pollution abatement, they reduce energy consumption, they increase the efficiency of raw material utilization, and clearly they're going to enhance PPG's ability to compete worldwide.

Thus, in every sector of the economy, the administration is seeking to carry out the President's directive to increase the efficiency with which we use, transport, and produce energy. Certainly our efforts would be aided if Congress would approve balanced energy legislation that the President could sign. Most of the energy efficiency provisions have already been agreed upon, and it is other issues that are the sticking points between the conferees. These agreed-upon measures include efficiency standards for new electric motors, fluorescent lamps, and commercial heating and air conditioning equipment; energy efficiency labeling for windows, light fixtures, and office equipment; increased support for stronger state and local efficiency standards for new buildings and for home energy ratings; expanded federal support for existing state and utility energy conservation programs – including the total elimination of federal taxes on utility rebates for improved efficiency; substantially increased funding authorizations for research, development, and demonstration of new technologies; and an increase on the tax-free limit on employer-provided incentives for use of public transit, while capping the tax-free value of parking that employers may provide to employees, thus trying to level the playing field between parking and mass transit benefits. Overall, we expect that if these new initiatives were to become law, they would produce important benefits for the nation, for manufacturers, and for consumers. Preliminary estimates project that they could reduce primary energy consumption by six to eleven percent by the year 2010, reduce the average household energy bill by about $60, and result in even larger energy savings for the nation as a whole.

Now, these savings won't be free. They will require billions of dollars of additional investment by manufacturers, utilities, and consumers, including the nation's largest energy consumer, the federal government. But this legislation has a number of provisions that will help facilitate these investments, including broader authority for energy performance contracting by federal agencies, and increased incentives for federal agencies to participate in utility-sponsored programs, or to seek their own budget authority for these investments. Together with DOE's existing programs and other federal initiatives, we think the efficiency provisions of this legislation could result in important environmental and energy security gains over the next ten to twenty years. And I remind the audience that this legislation would never have gotten this far without the President's leadership, because it was he who kicked all this off with his National Energy Strategy, released in February 1991.

In closing, I hope that my remarks have made clear the President's strong commitment to a more energy-efficient America. Under his leadership and direction we have launched important conservation initiatives in transportation, buildings, industrial processes, utilities, and federal sector energy consumption. What I have communicated to you are only the highlights of his far-reaching programs. There are some who are critical. I think this criticism is premised on one or both of two fallacious premises. The first premise is that maintenance of a robust domestic energy supply industry is somehow inconsistent or incompatible with a vigorous conservation program. The other premise is that the only meritorious conservation policy is one that places almost total reliance upon regulatory mandates, the old command and control approach. The President rejects both premises, because they are disproven by the historical record. Neither self-induced shortages of energy supplies, nor coercive regulatory mandates, has ever proven effective in the long-term. So way should we go down either path again? Why not try a new path, one that seeks a balance between supply and demand, and which is based on working partnerships between the public and private sectors? We firmly believe that this is the path that will enable us to fully utilize the greatest asset our nation possesses, the entrepreneurial spirit of free men and women.

Now I will concede that so far, our record in this area is still largely a record of inputs: money being invested, working partnerships being formed, plans and goals being established. Because we have chosen a new path we are, in a sense, just beyond the starting gate. But along this path, I firmly believe lie truly spectacular results, which, when combined with the responsible production of our own domestic energy resources, could provide America and the world with the clean, secure, and affordable energy supplies it will need in the 21st century. Many of you here today, along with the able assistance of the dedicated public servants at the Department of Energy, will make that possibility a reality.

INTERNATIONAL PERSPECTIVES ON ENERGY EFFICIENCY

Robert Fri, President, Resources for the Future.
Rohinton Aga, President, Thermax Limited (India).
Rogelio Covarrubias, Comision Nacional Para El Ahorro de Energio (Mexico).

Robert Fri: If you pick up anybody's energy forecast for world energy supply and demand over the next several decades, two things will jump out at you. One is the extraordinary growth in demand in the developing countries, which accompanies their economic development. The other is the extraordinary dominance of coal as the fuel of choice in supplying energy to meet that demand. Whether you like that depends on how pessimistic you are about the future. If you think that energy resources generally are in short supply, it's a disaster, particularly for the developing countries. If you're somewhat less pessimistic but nevertheless believe in the severity of the global warming problem, clearly it's too much coal. That throws you back on an uncertain dependence on technologies that are not yet in full flower. If you're an optimist, it's only money, because you can use any of these fuels, including coal, in a way that, save for global warming, is environmentally sound. But money, of course, is the scarcest of all resources, particularly in countries that don't have very much. So, no matter how optimistic or pessimistic you may be, energy conservation, energy efficiency, is critically important, for two reasons. It's cheap, and it's fast.

There are some stirrings of some programs to take advantage of energy efficiency practices and technologies worldwide. Our speakers this afternoon, representing some of the developing countries, will talk from their perspective about what is needed, and what role energy conservation and efficiency can play in solving their problems.

Rohinton Aga: I find that I am the only speaker here on international perspectives who comes from the private sector. Most of my friends are from the government, or are government-related. I wonder if this is a recognition by the United States Energy Association (USEA) of the winds of change in my own country. I will just tell you the highlights of the Indian energy scene.

India's per capita energy consumption when it became independent in 1947 was just 13 kilograms oil-equivalent. It has now gone up to 208 kilograms oil-equivalent. But even at that, it's well below the bottom of the list: China is about twice as much, Brazil is about four times as much, and the U.S. is about 40 times as much. On the power scene, in 1947 we were generating just over one thousand megawatts of power. Last year we generated over 65,000 megawatts. Out of this, 70% is thermal, 27% is hydro, and 3% is nuclear. The installed capacity and the energy generated have grown at an average of about 8% per year over the past eight years, except perhaps this past year, when it dropped dramatically. The reason for this decline was very clear: the government of India just does not have the funds for adding on new generating capacity. Over the last ten years the demand for power has outstripped supply. Last year the power shortage was estimated at around 8%, and the peaking shortage at around 18%. In some states, the peaking shortage has been as high as 30%. It is estimated that the power shortage, and the quality of power, contributed directly to loss of about 3% of our Gross National Product.

All the speakers today talked about the demand side of energy management. In India, the supply side bears some mention. Taking a 15-year look over the period from 1990 to 2005, it is estimated that the growth in demand for power would average around ten percent, as in fact it has done for the past several years. Growth of this magnitude sustained over a long period of time may be optimistic, but it has been achieved by South Korea, Thailand, Indonesia, Brazil, and Taiwan. A more realistic estimate talks of a 6.5% growth rate. On the other hand, the cost of generating one megawatt is estimated at between $1 and $1.2 million. To bridge the power gap over the next fifteen years, even assuming growth at the more realistic 6.5% annual rate, India would have to allocate something like 25% to 30% of its total investable resources to this one sector alone. This is totally unfeasible. This deficit will continue for the next ten years or more, and might perhaps be aggravated, unless resources can be mobilized outside of India, and simultaneously an action plan implemented for conserving energy. This is, I think, a challenge and an opportunity that this audience might like to reflect upon.

A word on new power generation. During the past 40 years, power generation, transmission, and distribution in India was a sacred cow, not to be defiled by profane private entrepreneurs. It was entirely the prerogative of the states. But last year the winds of change, emanating from the Siberian wastes, finally crossed the Himalayan Range, fanned a little by the International Monetary Fund and the World Bank. Lo and behold, the power sector – generation, distribution, and supply – has been thrown open to private enterprise, and without any national barriers. Here is an opportunity and a challenge. The government of India has published a glossy, well-worded booklet called "India's Energy Sector: The Scope for Private Participation," a copy of which I will leave with the USEA. Some salient features. Any entrepreneur, Indian or overseas, can set up a facility for generation, supply, and distribution of power. The entrepreneur will be allowed a debt-equity ratio of 4:1. Up to 100% equity participation is available to foreign investors. Import of capital equipment is allowed, with an import duty of 30%. License will be made available for a period of thirty years in the first instance, with a subsequent renewal of twenty years. The scheme provides for a two-part tariff, which will decide the cost at which power will be sold. This tariff talks about the fixed cost, with an escalation basis, and of variable cost, which will comprise primary fuel. All going well, the investor could be guaranteed a return of 16% of the total investment, post-tax. Repatriation of funds, both dividend and capital, are guaranteed. The government has advertised the projects which have been cleared or are under examination, of a value of about 21,000 megawatts. I am told that 24 memoranda of understanding have been signed, which include some US companies, like Southern Electric, Spectrum, and Enron. Considering that there is a general recession throughout the world in industries involved in the manufacture of power equipment, here is a wonderful opportunity for members of your association to participate in the scheme.

A second point on the supply side is the importance of life extension services for the Indian power sector. The plant load factors of thermal power stations in India have been uniformly low. Part of the reason is old power plants, which are more than 20 years old. Due to the normal process of aging, aggravated by highly corrosive Indian coal and poor maintenance efforts, this population of boilers is in immediate need of life extension. It is very reasonable to expect 20% improvement in availability of these boilers if a systematic life extension program is implemented. The second phase of this renovation and modernization program is underway in the current five-year plan. Two hundred boilers with a total capacity of 19,000 megawatts will be taken up for renovation. The estimated cost of this is around $500 million. The important point to note here is that the actual expenditure for life extension is around 10 million rupees per additional megawatt released, through innovation, compared to 30 to 35 million rupees which would be required for installation of fresh capacity. In the renovation program, there are three specific items that merit mention: turbines, control and instrumentation, and electrostatic precipitators, which are imported at the moment. I am sure that members of your association who were involved in the manufacture of these products and systems would find this of interest.

A third aspect on the supply side is cogeneration. Cogeneration, as you know, simply means simultaneous generation of thermal and electrical power. I am aware that the United States has done some very pioneering work in the area of cogeneration, both on the technology, and on developing suitable legal frameworks which make it possible for small producers to generate power and sell it to the utilities at a price at which everybody wins. It is estimated that some 60,000 megawatts of power are generated by small producers and then sold to utilities. In a path-breaking study done by the Bureau of Science and Technology of the U.S. Agency for International Development (USAID), it was estimated that the potential for cogeneration in India is around 10,000 megawatts. The corresponding equipment market was estimated at just under $20 billion. My own company, Thermax, has been working closely with USAID, and we've just completed a study of cogeneration at three specific sugar mills. With some reasonable assumptions, the rate of return on investment is attractive, with equity paybacks in the range of one to three years. One important but incidental advantage of cogeneration using biogas or other agro-wastes like husk is that it not only helps save scarce fossil fuel, but it also contains the CO_2 and SO_2 emissions that everybody is concerned about.

Let's talk about the demand side. Somebody once remarked that if all the money that was spent on the Alaska Pipeline, including cost overruns, had been spent on energy conservation measures, there would have been more saving of energy than will ever come through the pipeline. This remark in the Indian context bears considerable relevance. I would once again refer to this excellent publication by the Bureau of Science and Technology of USAID, called "Opportunities for Improving End-use Electricity Efficiency in India." The report analyzes the electricity consumed in India by different end-users, and finds that the largest end-users are electrical motors, 38%; agricultural pumps, 23%; lighting, 13%; and domestic appliances, 11%. The reports lists some 27 energy conservation measures which make possible large savings of kilowattage. It is estimated that if all the suggested measures are implemented, the forecasted electricity use and peak generating capacity requirements in 2005 would be reduced by about 20%. What's even more significant is that the average capital cost of these measures is only about 8000 rupees per kilowatt, about one-fourth the cost of installing a base-load power plant. The priority measures include improvements in the new and existing agricultural pump sets, metering of existing pump sets, the use of fluorescent instead of incandescent fixtures in new construction, improvement and maintenance practices in the industrial sector, and the development and promotion of efficient end-use equipment, including motors, variable-speed drives, electronic ballasts, domestic refrigerators, and the like.

The U.S. taxpayer might well ask why the U.S. government is financing efforts, through USAID, for energy conservation measures in India. I think there is a good reason. The potential on the demand side opens up a considerable opportunity for U.S. firms to supply equipment, components, and technology. The USAID report I mentioned talks of a market potential of around $17 to $24 billion.

A brief word on instrumentation and controls. There is a very big market for equipment and technology in the area of instrumentation and controls. My own company has made an entry in the area of real-time systems. We've done some market surveys, and the areas of some interest are the power sector (power distribution), steel, cement, refineries and petrochemicals, sugar, and railway electrification track monitoring. The demand projections for the next five years, for new plants and retrofit and modernization of existing plants, is estimated at about one and a half billion dollars, and growing at the rate of 15% per year. It's an area where equipment can be imported into India at concessional rates of duty if a project is related to conservation of energy.

A brief word on gas. India is a relatively new entrant in the gas scene, largely for ideological reasons. The government decided many years ago that gas should only be used as a feed stock and not as a fuel. Unfortunately for the government of India, the laws of physics do not obey national dictates. Wherever oil is tapped, gas is produced as a byproduct. It can only be flared or used. A colossal amount is being flared at the moment, something like one-third of the gas produced. We are now told that there is an ambitious multibillion dollar project to eliminate flaring, which has been prepared with the aid of the World Bank and the Asian Development Bank. Admittedly, gas as a fuel has less value-added than gas as a feed stock. But what is forgotten is that in the India exercise, the pollution abatement that would be possible using gas as fuel in replacement of coal or oil, particularly in highly polluted urban centers like Bombay, is significant. If the cost of pollution is recognized, gas as a fuel can come into proper focus.

A very brief mention of my company's mission in the area of energy, which is to conserve energy in industry through value-added components and systems which make it viable to reuse the waste heat from equipment and processes; to transfer heat from a fuel resource to an industrial application with the maximum efficiency; and to replace power for heating and cooling in industrial applications with a direct source of heat energy. We've done some pioneering activities in the area of conserving energy through replacing old boilers with highly efficient ones. We are leaders in the area of thermic fluid heating, where you can use heat at temperatures of around 350 degrees centigrade at atmospheric pressure to replace electrical heating. In our own units we have substituted for some 500 megawatts of electricity. We are also involved in fluid-bed combustion. I know in the West the focus here is largely for environmental reasons, but in India, where our coals have an ash content of around 40% and higher, the availability of agro-wastes like biogas and husk make fluidized-bed combustion a very viable alternative as an energy source. It bears mention that India produces some 70 million tons of paddy and 20% by weight of paddy husk can be used as fuel. This fuel resource has the oil-equivalent of five million tons of oil. Yet another area we have been involved in is industrial and comfort cooling. Traditionally this has been the preserve of reciprocating and centrifugal compression chillers, which are extremely electrically intensive equipment. Thermax has made it our mission to convert these compression systems to absorption chillers, using either oil or gas, steam, or waste heat. The electrical load relieved is on the order of 90%. Incidentally, vapor absorption machines also have the tremendous advantage of being environmentally friendly – they achieve refrigeration without using any CFC.

I would like to close with three brief points. One is the potential for joint research and development between Indian companies and their counterparts abroad. There is a wealth of talent available in India, at very low cost, which can be harnessed to good use through joint programs. Some of the areas worth looking at are: combustion of agro-waste fuels like biogas or rice husk to generate gas; pelletization of husk; liquid coal; mobile power stations using flared gas at sites to generate power; pressurized fluidized-bed combustion; and new turbine designs. These are areas where Indian scientists and engineers have some familiarity with the fuels and the markets. At the same time, some very fundamental work has been done in the U.S. A cross-fertilization of ideas and action plans can generate considerable synergy.

A second issue is the role of the energy service consultant. I believe there is a host of consultants in the U.S. who have built up a considerable expertise in identifying potential areas for conserving energy, developing feasibility programs, working out a payback, and most important, sharing costs and savings to provide credibility to the consultancy effort. I think there is good potential for joint ventures between energy consultants in the U.S. and those in India.

Finally I want to compliment one agency that is doing excellent work in energy conservation. That is the EPIC program of USAID. EPIC stands for Energy and Environmental Policies,

Innovation, and Commercialization. I am familiar with their work, and the thoroughness and the professionalism with which the programs are undertaken, from concept to commercialization. I only hope that we can look forward to their greater involvement and a larger thrust in the global effort toward conserving energy and preserving the environment, and in the process, generating a market in which manufacturers, designers, and consultants in the U.S. can participate in abundant measure.

Rogelio Covarrubias: The Mexican program of energy modernization grants a maximum priority to conservation and energy saving through an effort in which society at large participates. There is a global potential of energy saving calculated at more than 300,000 barrels of crude oil per day in the medium-term. It is essential to carry out an energy policy with realistic goals, and the industrial, energy, and transportation sectors are areas of high priority. In Mexico, the federal government should employ all the instruments and missions which, along with the permanence of a realistic policy of prices and tariffs, promotes the change of habits toward rational energy consumption.

The National Commission for Energy Efficiency gives advice about conservation and energy saving to entities of the federal administration, as well as to entities of state and local governments, and also private enterprises. The Commission is integrated by representatives of six ministries, the Mexico City government, the state-owned oil company (PEMEX), and the federal electricity commission. It's mission is to conceive ideas and promote strategies and guidelines on energy saving, to coordinate and establish agreements, as well as to promote the participation of public and private sectors in the development and application of regulations, and to diffuse technical and economic studies. All these actions are oriented to removing the barriers to allow market forces to work.

One market force barrier is the lack of knowledge and information. Government action is essential to create and maintain information systems that allow the identification of opportunities, the location of services and suppliers, the identification of sources of knowledge and experience, the identification of trends, and to provide bases for cooperation. Frequently the operation of market forces requires the breaking up of inertias. This can be achieved by promotion of actions, such as the identification of energy saving potential through energy audits and other studies, the support of high-risk initiatives, the dissemination of information, sponsoring and publicizing events such as conferences and seminars, and the design of incentive mechanisms, such as the step-handling of tariffs, accelerated depreciation of energy-saving equipment, etc.

Generally speaking, governments are large energy consumers, and they represent good opportunities for energy savings, and could be a very good example for the industrial and private sectors.

The Commission is promoting programs in seven areas. In each program, audits, studies, projects, diffusion, training, and research, as well as development actions are carried out. Some of them are conducted by the technical cooperation provided by different international organizations. The expected benefits from these programs will be reflected in improved industrial energy efficiency, in the decrease in primary energy and electrical energy demand, and in diminishing pollutants emitted into the atmosphere. There is an increase in advisory activity and technology transfer, and in the market related to energy conservation projects and associated services.

In the electric power industry, besides the need for new plants, it is possible to improve electrical energy generation efficiency through cost-effective cogeneration. Thermal power plant efficiency could be improved by 5% to 15% through investments with low payback periods. It is also convenient to improve maintenance and operating practices by the application of advanced technologies, such as on-line monitoring and predicting maintenance. Transmission losses can be reduced just by improving maintenance and circuit capacity. Installed capacity in Mexico is on the order of 27 gigawatts, and is expected to grow to 51 gigawatts in the year 2000. It is a national goal to reduce the burning of oil fuels, although we don't expect in the medium-range to replace all plants. The new plants will be designed bearing in mind high energy efficiency improvements. Also, demand-side management systems must be implemented in a very aggressive way.

In oil production, the government-owned national oil company, PEMEX, has been reorganized into four units: primary production, refining, petrochemicals, and gas distribution. Each unit will operate as an independent business unit, with its own board of directors and decision-making policies. The goals are framed in terms of total quality performance. The four units have to start a modernization campaign soon to mitigate the ecological impact due to exploration and production. It will require technical assistance, engineering equipment, and financing. In order to achieve important energy savings in the oil industry, intensive research activities are required. A real breakthrough in improving the energy efficiency of a refinery requires a substantial reduction of the temperatures at which the physical separation and chemical reactions take place.

In the industrial sector, a new area of opportunity is emerging due to the combined effects of increasing efforts at regulation, standardization, and certification of energy-efficient equipment and facilities, the regulations for environmental conservation, and the leveling of energy prices. This combination is spurring energy users to analyze the impact of energy costs in production, and forcing them to take initiatives to reduce energy intensity to comparable levels in international markets.

On the other hand, the production of electric power in Mexico was considered a strategic issue, and consequently it was a state monopoly by law. Because of this, a great amount of heat is lost to the environment through stacks, air and water coolers, and other pieces of equipment. In May 1991 the law was amended, and now it is possible to produce electricity, not only for home use, but for selling excess power to the utility. This has opened a wide opportunity for cogeneration, starting from the energy audit to third-party implementation, and possibly even for independent producers. The Commission has a preliminary potential cogeneration forecast on the order of 3.4 to 6 gigawatts: at least 500 large industries, with a potential of 2.5 to 4 gigawatts; 21,000 small- and medium-sized industries and services companies, which represents a potential of from 500-800 megawatts; and PEMEX refineries and petrochemical plants have a potential of from 800-1200 megawatts. Due to the high value cogeneration has in Mexico, the Commission has already implemented a series of actions to encourage industrialists, consultants, engineering firms, equipment vendors, and financial organizations to prepare feasibility studies and to develop demonstration projects, in order to trigger market forces in these emerging market opportunities.

The transport sector accounts for almost 40% of oil consumption. Daily gasoline consumption in Mexico City is in the vicinity of 20 million liters, i.e.,. 5.4 million gallons per day. This amount of gasoline, the relatively low speed of transit, and the topography of the geographical location of the city produce an environment heavily saturated with CO_2 and NO_X, and large amounts of ozone are generated nearly every day, which of course contributes to global warming. Fuel efficiency standards for our cars, trucks, and buses offer major opportunities for energy saving. Improved mass transportation systems can reduce our dependence upon energy-inefficient forms of transportation. It is possible to figure out a non-polluting urban transportation system by the use of electric cars, buses, and trains. This change will require the vehicles, the charging networks, and a number of related services.

There are opportunities here that will benefit all parties involved, from manufacturers and suppliers to the community in general. The expected benefits include: the reduction in consumption of primary fuels and electricity demand, which allows deferring capital investment to the energy supply sector; reduction of industrial end-use consumption, with the related improvement in productivity and competitiveness; and increasing the increment of the domestic end-use consumption to enhance economic welfare for Mexican families. The efficient use of primary fuels will reduce the greenhouse effect and other related ecological impacts, consequently improving environmental quality for urban populations. It opens market opportunities for consulting and technology transfer, and for energy conservation products and services, which can be negotiated through NAFTA (North American Free Trade Agreement) when it is signed by the three countries.

Michael Deland, Chairman of the White House Council on Environmental Quality, presents the Bush administration's strategy on energy use in 1992.

Mexico's Rogelio Covarrubias was one of several international representatives who spoke at the 1992 Energy Efficiency Forum as a result of the collaboration with the United States Energy Association.

Linda Stuntz, Deputy Secretary, Department of Energy, outlines the economic ramifications of energy efficiency in 1992.

Henri-Claude Bailly, Chairman and President, Hagler Bailly, Inc., has been a part of the Forum Planning Committee since 1992.

Fourth Annual

Energy Conservation Forum 1993

Fourth Annual Forum – 1993

ENERGY EFFICIENCY: THE NEW AGENDA – ECONOMIC GROWTH AND GLOBAL ENVIRONMENTAL PROTECTION

President Bill Clinton and his administration came into office with an explicit commitment to lead the U.S. toward becoming more energy efficient and more environmentally responsible. The early actions of the Clinton administration to implement this commitment inspired the theme of the 1993 USEA/Johnson Controls forum: "Energy Efficiency: The New Agenda – Economic Growth and Global Environmental Protection."

The forum took place in a heady atmosphere generated by the president's announcement two days earlier of a comprehensive Global Climate Change Action Plan, designed to reduce emissions of so-called greenhouse gases. (During the Climate Change announcement ceremony, President Clinton recognized Johnson Controls, Inc. as EPA's first Green Lights Ally of the Year.) The documents offered in this section represent a major portion of the proceedings of the forum.

There also was some discussion of one of President Bush's greatest legacies to energy efficiency: the Energy Policy Act of 1992. Among some of the groundbreaking changes provided for in this pivotal document include the deregulation of electric utilities, energy efficiency goals for all federal agencies and approval for the use of Energy Savings Performance Contracts.

In this section, the White House's Catherine Zoi describes the guiding principles of the Clinton administration's approach to energy and environmental issues. Newly named Secretary of Energy Hazel O'Leary gave a rousing and uplifting speech asserting that the concept of energy efficiency has been transformed from a wallflower into the "belle of the ball" in public consciousness and public policy.

Other speeches in this section include a panel of analysts discussing the specifics of implementing energy efficiency measures in the U.S., including the concept of public/private partnerships, trends in the national energy marketplace, the promise of new technologies, and obstacles to advancing energy efficiency programs. You'll also witness a show of bipartisan cooperation as Democratic Senator John Glenn presented Republican Governor Tommy Thompson of Wisconsin with the domestic energy efficiency award.

Finally, Gerhard Ott of the World Energy Council delivered a sobering summation of the WEC's important report *Energy for Tomorrow's Future* on the immense challenges of delivering energy, efficiency, environmental improvement to people around the globe. Ott noted that developing countries consume about one-third of the world's energy, but by 2020, they will consume approximately 55% of the world's energy. It will require a $30 trillion investment to meet the energy needs of these nations.

These words certainly were timely for members of the 50 people representing 17 developing countries as part of the USEA Foreign Study Tour; but in reality, it is the rest of the world who also must pay attention to these concerns. Our horizon is global, but the theme is simple: Leaders from the entire world must work together to reduce our collective energy consumption.

THE GUIDING PRINCIPLES OF THE CLINTON ADMINISTRATION'S ENERGY AND ENVIRONMENTAL AGENDA

Catherine Zoi, Chief of Staff and Deputy to the Director, White House Office on Environmental Policy.

I thought I would spend most of my time giving you a snapshot of the guiding principles of the White House and how energy conservation and energy efficiency fit into the way we work.

What is guiding this new and energetic administration? Bill Clinton has a number of guiding principles, and one of them is: **making change our friend**. There are global economic forces that do not make much sense to try to combat, but to instead embrace them. If the U.S. is going to enter the 21st century as a continued world leader and economic leader, and maintain our strength, we need to embrace change.

The President is committed to **maintaining our strong position in the world economy**, and that continues to make energy conservation very, very important. He will **not shrink from difficult problems**. I will illustrate some things that the Clinton administration has already done in the environmental arena that demonstrate that.

One of our cornerstones so far has been the **reinventing government initiative**. We know that the federal government is a huge force, but some of the institutions in place make it difficult for government officials like Mark Ginsberg to do his job implementing energy-saving programs throughout the government. One of the things that Al Gore has taken on under the President's direction is to streamline government to make it easier for the federal agencies to embrace energy efficiency so that the paperwork demands are not so onerous that projects aren't undertaken. Another guiding principle that you have heard in President Clinton's remarks on the climate change plan, on his health care plan, and when he talks about NAFTA, is the concept of **shared responsibility**. The government can help, and the government is there to help, but, in addition, it is up to each and every one of us to do what is right to position this country for the 21st century, so that people can have satisfied lives and the world will be safe for our children's children.

What these general principles translate into for the environmental agenda is the operative principle **that the environment and the economy function together**. For the past 12 years a lot of "environment versus economy" rhetoric was spouted by politicians and some representatives of industry. The Clinton administration orientation is fundamentally different. We see demonstrations at every turn that environment and economy work together, and I know I don't have to make too tough a case for this group, because this audience knows it and breathes it every day.

There are many people on President Clinton's team who know it too. Al Gore has a long history of being extremely committed to these principles, and he is there at every turn, supporting initiatives that marry environment and economy. In the selection of the cabinet members, Hazel O'Leary has already been a great leader at the Department of Energy (DOE). Carol Browner of the Environmental Protection Agency (EPA) has a terrific track record in Florida for seeing at the state level how the environment and the economy function together. Bruce Babbitt, Jim Baker at NOAA, Jack Gibbons the science advisor and the budget team are all strongly committed to making the environment and economy work together, and we are going to demonstrate that in everything this administration does.

So where does OEP come in? OEP is the Office on Environmental Policy, where I reside. We are a new creation of this administration, as a commitment to environment as a key organizing principle in all the initiatives the administration undertakes. We are comprised of a dozen people in total, and we staff the President on key environmental issues and organize interagency activities, which have historically always been cumbersome and unwieldy. I'm proud to say that we have made a lot of progress in eight short months, and I would like to quickly delineate what we have been able to do.

- **The Forest Plan** – Some of you may have read about the "jobs versus the Spotted Owl" dispute. Again, on the principle of not shrinking from difficult problems, Bill Clinton in his first

100 days in office, held a summit in the Pacific Northwest and brought people to the conference table, out of the courtroom, and said "let's get this done." Six weeks after that the administration published a Forest Management Plan that considers the jobs, that gives economic support to the loggers who might lose their jobs, so that we can maintain and protect our old growth forests, which is part of out national heritage.

- **The Wetlands Policy** – Wetlands historically have been an extremely divisive issue, with developers being pitted against environmentalists. This summer we published a Wetlands Policy that would again reduce red tape but still maintain the principle of no net loss in our wetlands.
- The President signed the **Biodiversity Treaty**, a treaty that George Bush refused to sign because he said that it would hurt business. Once again, this administration worked *with* business to change some side agreements associated with the treaty, so that business was comfortable and so that our biodiversity could be protected internationally.
- The President established the **President's Council on Sustainable Development (PCSD)**, a group of 25 innovative, creative, leading thinkers in this area that will, over a two-year period, come up with recommendations on how to implement sustainable development, so that within five years we will not need to utter the phrase "Sustainable Development," because *all* development will indeed be sustainable.
- The President has also signed a number of **executive orders**, including one yesterday on the procurement of environmentally preferable and recycled goods. This executive order is being hailed by environmentalists for the first time creating large markets for recycled products. Mayors of many large and small cities are hailing it because, while they've zhad aggressive recycling programs in their cities, there was no market for the material that had been collected. The executive order will also create jobs in cities, often jobs for people that are somewhat difficult to employ. The administration has also issued executive orders on alternate-fuel vehicles, ozone substitutes, toxic emissions for federal facilities, and energy-efficient computers.

I am very proud of this track record so far, and this administration will continue to work tirelessly to advance this agenda, keeping in mind those principles; embracing change, positioning the U.S. for a world economy, addressing difficult issues head-on, streamlining the federal government, and sharing responsibilities among all the players.

So what's the big news most relevant to this audience? Probably the biggest news is the Climate Change Plan. I would like to quickly go through some of the guiding principles there. First of all, the President made a commitment on Earth Day to reduce greenhouse gas emissions to 1990 levels by the year 2000. Make no mistake, George Bush wouldn't make that commitment – he didn't make that commitment. That's a target with a timetable; those were the two "T" words that the previous administration wouldn't embrace. The plan has about 50 different initiatives, and the reason I emphasize that is that this audience may all know, there is no silver-bullet solution to climate change. It is pick-and-shovel work, and this administration is intending to get this job done. The plan includes all significant greenhouse gases; carbon dioxide, methane, nitrous oxide, HFCs, and others. It takes measures in all sectors of the economy where greenhouse gas emissions occur.

Climate change is an economy-wide problem, so President Clinton is proposing economy-wide solutions. The plan fosters partnerships with business, which are so important, through focused government guidance and flexible approaches, to produce cost-effective emission reduction. The plan stimulates investments in the technologies of the future, and strengthens the American position in the global environmental technology marketplace.

The next point is really important to me, having come to the White House from an EPA program that was funded on a shoestring budget. This Climate Change Plan is backed up with real federal resources. The administration is committing $1.9 billion in new and redirected funding between 1994 [and 2000] to make sure the action plan is real. It reduces the deficit through a couple of new policies; one of them will allow commuters the option of cashing out employer-paid parking; the other permits private development at existing federal hydroelectric facilities in exchange for lease payments. These reforms will raise about $2.7 billion between 1994 and 2000. It leverages over $60 billion in private investment. Sixty billion dollars in private investment turns into $260 billion in revenues saved from energy bills between now and 2010.

The Climate Change Plan creates new jobs. Again I am preaching to the choir in this forum, because the energy efficiency industry knows that it's more job-enhancing and labor-intensive to do energy efficiency upgrades than to build new power plants. A lot of America does not know that. This plan, because of its reliance on energy efficiency, does create new jobs.

The Climate Change Plan coordinates multiple programs across federal agencies to enhance their effectiveness. There used to be much divisiveness between government departments – particularly EPA and DOE – and an unwillingness to work together. I can tell you, because I was part of those debates in the previous administration as we were trying to get George Bush to be aggressive on climate change. There is genuinely a new spirit of cooperation. The plan has programs that are designed and crafted with DOE and EPA working very, very closely together, with EPA and the Department of Agriculture (USDA) working very closely together, with DOE and the Department of Transportation (DOT) working very closely together, in a spirit of cooperation from the top. This plan is designed for rapid and aggressive implementation. Some are skeptical because the approach of the plan is largely voluntary. But, as Secretary O'Leary points out, "voluntary" isn't a dirty word. This plan is designed so that work can begin on it today. New legislation is not required for the most part. The plan builds on existing successes in the existing institutions, and expands them more widely across the economy. At the same time, the administration is going to actively review and monitor progress to ensure that it is on track.

The President is committed to ensuring that his target is met within his timetable. Which means that if in two years we are not on track, if programs that we thought would bear fruit are not bearing fruit, we will reallocate resources, or we may need to go and do things that are more difficult from a regulatory or legislative standpoint. The administration will actively monitor and review, both from the emissions side and from the programs side.

A White House team has also been established to develop strategies for long-term emissions reduction. Bill Clinton took a big first step two days ago, and that's what it is: it's a critical first step *and* it's a giant leap. Climate change is inherently a long-term problem so it will necessitate a long-term solution, and the administration recognizes that and bring in the Science Advisor and the National Economic Council to figure out how to address that in a long-term way.

I'm really proud of this plan, it has been a Herculean effort and the right effort at the right time. How does that translate into what this industry does? The analysis of the plan suggests that approximately 50% of the potential greenhouse gas reductions are from energy efficiency improvements. About 35% of those reductions are estimated to come from the residential sector, 23% from the commercial sector, and 41% from the industrial sector. Again, these reductions are economy-wide, with different programs in each of those areas. There are robust standards in the plan with aggressive market pool, and I am confident that the result can and will be real. As I pointed out, my experience with the green programs at EPA demonstrate that this approach can work, and not only can it work, it can be much more successful than other approaches.

I remember hearing a story after the EPA launched the Energy Star Computer Program in 1991. Energy Star is a partnership between the EPA and manufacturers of computers to put energy efficiency technologies into their computers. Basically what Energy Star computers do is go to sleep when they're not being used and, therefore, save great amounts of energy. A friend of mine approached a representative of Hewlett Packard after we made the announcement (Hewlett Packard was one of our charter partners in Energy Star) and said "When did you guys start thinking about energy efficiency in your computers?" The HP representative said, "Well, you know, when EPA came to talk to us about it a few months ago." There is opportunity like this across many, many technologies. And now we've got virtually 100% of the computer manufacturers making Energy Star products, actively marketing them. The federal government is committed to only buy Energy Star computers. There is an enormous amount of opportunity here and in lots of technology areas, and that's where this plan and its administration is headed.

So what's next? Well, implementation of the action plan: it's not a "set- and-forget" idea. We may want to start with upgrading the 6,500 hospitals that someone mentioned. (The President really liked that part of the health care plan!) Can you imagine that? That energy efficiency savings is worth about $1 billion. If this plan means that we need to upgrade 300,000 buildings, then let's do it. Let's do it together. Six months from now, the administration is committed to holding a major White House conference to see where we are. Program managers from the Department of Energy, from EPA, from USDA, from DOT, will tell you about the progress and the role of their programs. The White House will invite private-sector stakeholders, and public citizens who will be the recipients of energy efficiency upgrades, to come together to say "where are we, what help do we need, what momentum to we need to continue to build?" That will happen six months from now, on the anniversary of President Clinton's announcement.

This Climate Change Plan means big business growth for this audience. It can mean big savings for your customers, and big, big, big, benefits for the environment. We are really committed to working with the industry on this.

What else is the administration going to be doing in energy efficiency? We have got an executive order on energy efficiency that will be issued in several months. Bill Clinton will not sign an executive order that is not a clear directive above and beyond what is already on the government books. He will not issue an executive order simply to have an executive order on energy efficiency. The administration will be aggressive, practical, and give business and the federal government a sensible blueprint for proceeding to build some momentum from what is already in the Energy Policy Act, which is a fabulous start. If the President needs to give government officials more guidance on how to pursue energy efficiency in their areas, then that will be incorporated into the executive order. The White House is working really closely with Mark Ginsberg of the Federal Energy Management Program to craft an executive order that will be helpful to federal agencies, rather than an additional barrier to improved efficiency.

There is also an active project that many of you are familiar with: the Greening of the White House. Again, the President and Vice President want to make the White House complex a showcase of energy and environmental efficiency. I know Mr. Ginsberg is having to spend a lot of time on the Greening of the White

House project. To date, DOE has provided an audit of our energy operations at the White House, and that audit is being reviewed internally to ascertain what can be done this budget year and next budget year. To do the job right, Congress may need to appropriate additional money so that investments can be made that will pay back in three to five years. That is good for taxpayers and, if necessary, the White House will go to Congress and ask for that money.

This is a quick summary of what is going on in the administration that might be of interest to this forum. We need your help, we need your leadership, we need you to communicate back to us where other things like this, that are consistent with Bill Clinton's guiding principles, will help get this job done. I thank you so much for the opportunity, I'm delighted to be here, and again, thank you for all of your work and your efforts.

ENERGY EFFICIENCY: THE BELLE OF THE BALL

Hazel O'Leary, Secretary of Energy, U.S. Department of Energy (1993-1997).

I've got a simple question to ask this group, with whom I think I have so very much in common. Energy efficiency: is it a wallflower or the belle of the ball? Now I think among this group, we probably already have the answer, but let me focus on this just a little bit more. First of all, some of you may know that in the United States it's Energy Awareness Month. A little known fact is that today the District of Columbia will sign up as a Clean City, indicating its desire to have the government focus on D.C.'s need to draw a market for alternative-fuel vehicles. Yesterday we were out on the lawn of the White House and President Clinton announced our global Climate change Action Plan. It seems to me the question is answered: energy efficiency is the belle of the ball.

Now, having understood why the belle is getting so much attention, let's see what generally happens at the ball, and my question would be "Why dance?" Why dance, indeed. Well, for several reasons. We dance because this is an administration that's focused on issues involving the environment, energy, and energy security, which also means national security. We recognize the very strong ties between the environment, the economy, energy efficiency, and our national security. We've linked energy and the environment in a way I believe that no other administration has. And why would that be? Again, the "why dance?" piece. Well, first of all there have been public imperatives heard by an administration that I believe came to office on the environmental ethic, among many other principles of our administration. Received resoundingly by responsible companies who understand, that really do get it, that without the clear commitment to sustaining the environment there can be no trust, no confidence in an entity that purports to do business in this country and internationally. With respect to the energy and the environment link, if we do this correctly then we'll also have an opportunity to address the deficit, because what we know about energy efficiency is, it's more labor intensive, it's generally cheaper than bringing on the next increment of power. It answers another question with respect to security. It's already been projected that during my watch at the Department of Energy, we're going to top 50% in terms of imports, and that is unacceptable to the President of the United States, the Vice President, and to me personally.

To continue my theme, when it gets a little overbearing just start to groan and I'll stop. So we know we want to dance, and if this is one of those nice old fashioned tea dances, then we've got to fill up the dance card. How do we do that? Well, you have to have a plan, my father told me that, before you start anything, get a plan. We now call that by the highfalutin term of strategy. Quite frankly, that was laid in place by the Congress last fall when they passed the National Energy Act of 1992. And some of us may argue that a piece of it was left out, but certainly that leg of the stool, called energy efficiency and alternative energy, is well in place, and the challenge of my colleagues and I at the Department of Energy is not what to do, but in what priority and what order to do it, and how to make certain we're getting the right funding for the programs that give us the biggest bang for the buck, both in terms of the economic stimulation and in terms of energy conserved and of course, environmental improvement. The funds are committed and perhaps most importantly, the Department of Energy is committed to finding another $222 million to support the initiative for global climate change, because 70% of the programs coming out of the energy sector rest with the Department of Energy.

Have the partners been identified? I think so, I really do think so. When I was back in the business of regulating the energy sector, we started our relationship with the utility sector in the United States. And I don't know that they liked it a whole lot, we called it 'camel's nose under the tent,' you know what happens after that, but we had the good sense at the federal level to shift most of that responsibility for forcing regulation that would encourage energy efficiency to the state level where it made a lot more sense. And not only have we got the utilities as our partners, but we've got the state commissioners as our partners. We've got state energy offices as partners, and we also have the broader industrial users as partners, because as you begin to use your utility sector to focus on energy use you end up with the large energy users, your focus has got to be on the industrial user. From time to time the commercial user if it's big enough, and the households, because that matters too. So, the partners are in place.

Are the users aware? Some, most, the big sophisticated users are because they've already seen that the bottom line in terms of profit to the company, the bottom line in terms of what you pay out if you happen to be a household, might be positively impacted if energy conservation techniques can be applied. So I think we've got the users aware, but I will admit to you that we need to be working in a partnership to make them more aware.

All right, you can groan anytime you want. This one is called "mastering the steps." What you're seeing here is a very powerful display of the kind of work that's going on, specifically in the Department of Energy, in energy efficiency projects. And so we've got our utility group, we've got biomass projects, competing now for joint venture funding, we've got existing or proposed wind power projects throughout the United States, and we've got targeted drilling for geothermal initiatives. I want to leave this by simply letting you understand that the commitment in the Department of Energy is not merely for energy efficiency, as one would think about conservation in the old days, but with a real push for the alternatives, because we know not only are there marketplaces there for alternative energy use, about which I'll talk in a minute, but it certainly is a more benign energy source when it's cost-effective.

How do we take that initiative? And this one is really bad, I do apologize for this. We're calling it "the makeover," because having become the belle of the ball you suddenly thought about 'Well, she gets the new dress, the new hairdo' and you've got it. And in many ways that's happened in the Department of Energy. When I arrived we were still calling it "conservation." Kind of old style, not really recognizing the economic tie between saving energy and saving money, and more importantly, not pulling in the environmental component. This part is as busy as it can be, both nationally and internationally in the areas of energy efficiency. I talked to you about Clean Cities a moment ago. Almost every week Frank Stewart and his people have been in another major city in the United States, visiting with the mayor who has said "My God, we want to be a Clean City." When he's not answering the phones because he's off in some exotic place doing that, I'm answering the phone with yet another mayor saying "Hey, I think we could qualify to be a Clean City." In the next few weeks we'll hit Boston and New York and Chicago, until we've identified 38 Clean Cities for this year. And we'll use the government funding to force an opening marketplace for alternative-fuel vehicles, but we'll only go into cities where the ethic exists and where there is the beginning of an infrastructure. In the old days, when I had the old conservation and environment job, we were mostly being pushed into this work and begging for a budget, but that's no longer the case.

I want to stop for a minute and talk about the global climate change initiative of this administration. Press coverage has been broad, I opened the clips this morning and saw more of us then I ever expected, and I think that indicates the interest of the public in this project and in this commitment of our nation. There is a change in point of view from the Bush administration to the Clinton administration. They were wondering if and could and should, is the science there. We've taken a totally different position. We've said we don't want to quarrel anymore about the science. Meeting the goals of the global climate challenge internationally, which is to sustain emissions at the 1990 level through the 21st century, can be done basically with energy efficiency. In the energy sector 70% of the gain to reduce emissions comes from energy efficiency. The other 30% comes from fuel switching, and the beauty of this piece is that we can do it without that old command and control approach, for two reasons. It's very simple. First of all, the states have kind of got a hold on industry through their utility regulation. More important, the push is there from the public. I had the privilege about two years ago being a part of a utility that signed up for the Valdez principles and actually came up with an environmental policy. Well that's old hat today, any company worth its salt is doing something like Valdez. My point is that we can make the goal for the 21st century by simply relying on letting the private sector do what it must already do and quantifying emissions reduced. Why would we want to do it that way? Well, it's more cost-effective, it's flexible, it allows the business planner to do what makes the most sense in his or her region of the country, depending upon energy source and fuel mix and the needs and demands of customers. It makes a whole lot of sense to me.

I want to talk about the bucks for just a minute more. There is a broad marketplace out there for this technology. There is a big marketplace existing south of the border. For energy conservation techniques, retrofits at power plants, retrofits to large manufacturing concerns, some have quantified that the marketplace internationally approaches over $200 billion. Sounds to me like the belle of the ball has been entirely made over, and I know that everybody wants to play. How do I know that? I can't go to a conference, a meeting, an international or border gathering where they're not more people crowding around with that deal in the back pocket, and perhaps more importantly on these issues involving energy efficiency, I see the big money folks around. The investment bankers are there. That says to me that not only have people decided that the risk is worth going after, but the reward in terms of the margin of profit is there in this business. It makes sense all around to me.

DOE is conducting other energy efficiency dances Energy analysis and diagnostic centers, which are places where business folk can come who are not from large companies that have their own energy experts, as many of the companies represented in the room do, but small- and medium-sized businesses who could improve their bottom line and make a real contribution if they knew what technology to apply. We've established these centers and some of them have been so successful that they're already on the Hill testifying not just to the good news of the fact that they exist, but they actually have brought along business people with them who report, "I have worked with this center and saved myself a ton of money and been able to hire on new people." That's a real win for me. Energy audits are key. And the state energy offices we love to count because again, it points up the partnership, and also we've logged the nine Clean Cities.

Now, since I've talked about all those activities I thought it might be a good idea to give you some sense that activity does, in fact, lead to results. What this chart shows is the total savings so far in terms of dollars saved that we can quantify. Recall that we didn't know how to quantify ten years ago, or not how to do it well. Now we are focusing not only on the activities, but on the quantification of the results in terms of the energy saved, dollars saved, and pollution prevented, so we'll be talking to the public from the Department of Energy in a different vein.

There are other dances we can do, and one of them is the hat dance. This permits me to again focus on the new markets and it gives me a chance to do my NAFTA (North American Free Trade Agreement) lift, which I'm doing at least through the rest of the month until we win on the Hill. I've talked about south of the border and the opportunities for energy-efficient technologies and businesses. I was over on the south lawn yesterday in the rain and the muck with any number of companies who were selling control equipment, that helps large manufacturers through computer assistance save energy and send the right dispatches so that we are using equipment at the appropriate time. There must have been over 100 businesses on the lawn in the rain displaying their products, and the amazing piece of it to me was how many of those projects tied directly to energy efficiency, be it power plant retrofit or applications in the manufacturing process. The point I want make here is, not only does NAFTA open up new jobs for the nation and Mexico, but it opens up all of South America, which all of us understand is the growing market for opportunity in terms of global markets. Our belief is that if we take Mexico then we can march right straight south and take every other country that has now begun the effort at privatization, and perhaps more importantly, provides vast markets for decentralized energy applications. What does that mean? It means good business opportunities for small- and medium-sized companies where many of the jobs are going to come from, and perhaps in isolated villages and hamlets permits the best opportunity to try out some of our alternative energy technologies, and once again, fires up the marketplace and creates jobs back at home. That's just such an easy win I'm amazed that we're still talking about it. But we will talk about it until the very end.

I've hit upon the energy efficiency piece, but I would be remiss if I didn't in this audience that's going to talk about energy generally, tell you that for applications in generation, for both electric utilities and production and distribution of gas, the market is even broader. The important piece of that market is to sell U.S. technology, which is more energy efficient and environmentally correct. The same pertains for the Far East. This summer we sent a marketing crew off to China and southeast Asia. I was at one of these huge dinners that everybody attends, and the Chinese Ambassador to the United States was at the dinner party, and the way I'm describing it is that he practically sat in my lap. He said "Oh, Secretary O'Leary, we're so excited because we want U.S. technology in China. Both for energy generation and for energy efficiency." Those markets are there. Now if we don't get them, someone else will. And I'm saying we have the best technology.

Now, this allows me to tell my Russia story. I was in Russia just two weeks ago, because our administration has made a commitment to the Russian government that we will work on trade issues because we want to further encourage democratization, and quite frankly, we want those markets. The market for energy efficiency, not just oil and gas. In energy efficiency alone, Russia offers an opportunity of about two billion dollars just this year with projects we know will be ongoing and up and running by next year, and the market will grow. Now, I'll point out something that to me was mind boggling. The prime minister of Russia, Victor Chernomyrdin, was the President of GasProm, and gas perhaps, is the energy resource that the Russian Federation counts on the most, because it can see markets west. The prime minister, quite without a lot of prompting said, "I understand that energy efficiency is the answer for the Russian Federation." And then he went through the analysis. He said it's cheaper, much, much, cheaper than producing that next increment. Well, why is he so interested? Quite simply, because if they can conserve in Russia, then there's more gas they can sell to the west. That's fine with us, do you know why? For two reasons; it makes sense in terms of the environment. The United States has the technology to bring into Russia, the first project up and running will be to relamp all of Moscow, saving 50% in their energy costs just by that one act. And the second, which is something that will help for anyone who's driven into Moscow from that ring road from the airport, is to light the ring road, which has never been lit and which is a danger to anybody who pretends they want to use it without their heart in their throat after sundown.

There are other opportunities coming out of Russia and Eastern Europe. Trade missions with clean coal technology, trade missions with energy-efficient technology into Poland and Czechoslovakia, and anybody who's ever been there on a winter's day and seen the pollution understands that we've got to scrub and retrofit and then we've got to begin to get on it with energy efficiency. In these two areas of the country I will tell you also that if you need a hotel room you better book it six months in advance, because anybody with a contract, an idea, a gizmo, is there selling. So several things are required. One, highly principled business people going over, serious and with products that work, and that sounds to me like folks from the United States of America, and it also sounds to me like people who have been doing this business internationally and doing it well. There's a marketplace there and we're all going to be struggling to get it.

We've talked about this belle of the ball, we've talked about this partnership. I want to point out something to you. In terms of the focus of the Congress of the United States working with almost any administration, there are some solid alliances and blocs that have existed since the beginning of time. I know them well because I've done the dance a lot. Coal producing states and their leaderships, oil and gas producing states and their leaderships, and perhaps most importantly if you look at that big picture, those that produce both, have traditionally been the power base in the Congress when energy policy has been set. I would submit to you that maybe we want to change that. We want to change that a lot. What we've attempted to show on this map are the areas of the country where there are principal leaders who speak out in support of energy efficiency. Now I could get a little bitter here because I'm just about finished with my struggle for appropriations for the Department of Energy's budget. I will tell you that the energy efficiency budget will be increased, I believe, by 18.9%. And I'll tell you we started off with about approximately 33% on average. Am I satisfied, no. Why am I not satisfied? Because in spite of the support and good will coming from those states and the leaders from those states in the Congress who are proponents or champions of energy efficiency, we need to do a better job building up those advocates and expanding the numbers. We've got some Congressional sponsors of this conference who are strong advocates of energy efficiency. Ed Towns, George Brown, Vic Fazio, Ed Markey, Herb Kohl, Pat Schroeder, and Phil Sharp. But we can broaden that constituency. If we were called upon today to mention the names of the people who have championed what we used to call the hard path energy, we wouldn't have to shake our heads, or scratch our heads a lot, we would just spit them out. Our goal ought to be in the United States, to make sure everyone in the nation can spit out the names of the leadership around energy efficiency and alternative energy, and that the group be so large that there not be enough time to read it off, to mention them all.

So I come to the end of this piece by saying that it's time to strike up the band. And how do we do that? We do that first of all, by this continued working on the solution of the barriers that exist, which perhaps means we have to do a little bit about how the taxes favor one energy source against the other, and we can work on that in the coming year with the next energy strategy. We need to talk about dollars and cents and bottom line market successes, and that's to one audience, and then we need to talk about energy saved, because that's another audience, and we need to talk about positive impact on the environment.

But we have to talk all the games in terms of our market successes. And finally, we've got to expand that base of support. Now, I'm really either going to apologize for this one or you're going to think I'm absolutely brilliant. Some years ago there was a movie, and I think Joann Woodward was the heroine and the setting was an institution, a mental institution, and at one point she was talking about being at a resort and dancing with a marvelously charming, absolutely fascinating and brilliant man, who in conversation as they were dancing said 'You know I'm only here for the weekend' and her response was 'I'm dancing as fast I can.' And my point here is that we perhaps have a little longer than the weekend, but at this moment when history and economics and concern for the environment are converging to generate increased awareness of the significance of energy efficiency, we have a great opportunity, but not necessarily a lot of time. We've got to use that window of opportunity now while it's there.

IMPLEMENTING ENERGY EFFICIENCY IN THE U.S.: MARKETS, PARTNERSHIPS, TECHNOLOGY, AND ENVIRONMENTAL PROTECTION

Dennis Wamsted, Executive Editor, The Energy Daily.
Michael German, Senior Vice President, American Gas Association.
John Hoffman, Director, Global Change Division, U.S. Environmental Protection Agency.
William Nitze, President, The Alliance To Save Energy.
Gerald Thorpe, Director, Maryland Energy Administration.

Dennis Wamsted: Secretary of Energy Hazel O'Leary briefly talked about the importance of money and daily budgets, and I think that I would be remiss if I didn't remind you of the importance of that. The daily agency budget still does not give very much weight to energy conservation issues, and that's probably something that you all should focus on. The need for money will be particularly important in the next few years as the department tries to implement the Climate Change Action Plan that was announced on Tuesday. The secretary noted that the department is going to need $222 million to do each share of this plan, but what the secretary didn't say is that there isn't going to be any new money for that plan; that's going to have to come from the existing budget, and it's going to be up to you all to get her and the rest of the department to take that money from less desirable sources, or less desirable parts of the budget, and focus it on energy conservation and efficiency issues.

We'll be starting today with someone that I've known quite well for the past seven years. We refer to Mr. German as the sort of statistical guru in the Energy Department community. Whenever you need to know something about natural gas, you can call him up and he'll have the answer. He will be speaking to us first this morning and his official title is Senior Vice President of the American Gas Association (AGA), where he directs all their planning, analysis, and their statistical organization. Before coming to AGA, Mr. German worked in the Department of Energy.

Michael German: We worked together, several of us, a year or so ago on a document called the "Alternate Energy Future," and it was co-authored by the Alliance to Save Energy, AGA, and the Solar Energy Industry Association. The Global Climate Change Division of the EPA was a peer reviewer and very influential in the document itself. So I think the gas industry, and industry in general, is committed to energy efficiency and energy efficiency strategies.

That should raise a question as to why. Why should energy-related industries like efficiency? The bottom line is, they shouldn't, and basically don't. But there's a difference between like and need. And energy industries need efficiency. They need efficiency to compete in the U.S. and global marketplaces. Without it they will fail.

I represent the natural gas industry; we compete against oil in every market, from residential and commercial, to industrial, power generation, and, increasingly, vehicles. We compete against electricity, our principal competitors in the residential, commercial, and industrial sectors. We compete against coal in boilers, from commercial right through power plant-size. And either we're going to get out there and promote more efficient gas equipment – and we're talking about from well head through the burner tip, through end-use equipment – or we will lose market share and we will eventually die. Either you move forward with more efficient equipment, which lowers the cost for somebody using your equipment and makes your production and transmission infrastructure more efficient, or your market share will contract.

That is a fundamental bottom-line, and that is why it is in the gas industry's interest to be aligned with the Global Climate Change Division of EPA and with the Alliance to Save Energy. It is out of self-interest that we are aligned. But it is a self-interest that will result in the stabilization of CO_2 and much lower energy balances than would have existed otherwise.

Now, what do we see for U.S. energy balances and natural gas? Energy efficiency will result in the energy industry in this country being mature. By mature, I mean very slow or no growth. Analyses to the contrary are fundamentally flawed, because they don't recognize technological advances spurred by competitive pressures. We talk about stabilizing CO_2 emissions at

1990 levels, in 2000; in 1993, CO_2 levels will be roughly the same as they were in 1990. In a sense, we have been stable for three years, and if we can continue to progress along efficiency lines, we will remain stable. The gas industry currently has a 25% market share of total U.S. energy consumption. We see ourselves as a growth industry. If we are going to grow in a mature market dominated by new technology, it is incumbent upon us to get better and new equipment into the marketplace, and it will only get into the marketplace if it's more efficient. And this goes for entering the vehicle market, which is a major goal of the natural gas industry, competing for the electrical generation market with commercial and industrial cogenerators, or up through utility power generation. In addition to efficiency, there are things in the action plan relative to Clean Air Act implementation, which are extremely important to the gas industry, to maintaining and increasing market share in the residential and commercial sectors.

When it comes down to it, that is the guts of the gas market. We've done very well – we are converting 300,000 homes per year from other fuels to natural gas. Some of them are electric homes, most of them are oil homes. We're hooking up 1 million new homes per year. The highest market share for gas, in 1992, was 63% in the residential sector. The thrust behind this is total cycle efficiency. The electrics will continually challenge us with new equipment, both on the production side and on the end-use side. It is up to us to meet that challenge with more efficient equipment, and next year the gas heat pump will be broadly test-marketed across the U.S., equipment that will again leap efficiencies 50% on the heating side. I'm not going to touch on efficiency on the supply side, but we have joined hands with the Global Climate Change Division and John Hoffman in the Methane Star program, the natural gas star program, with the goal of making more efficient the transmission/distribution infrastructure and cutting back on leakage to the atmosphere of methane.

I'd like to quickly conclude with what I see as the four major energy/environmental issues of this year, next year, and the year after. The ones that will impact efficiency and market share greatest in this country and probably have worldwide implications. The number one issue, and by far the most important, is Clean Air Act implementation. It gets very little press, but it's going on today and it makes a big difference on energy efficiency, energy use overall, market share, and CO_2 emissions. The goal on the table today is attaining clean air in the urban centers throughout the northeast, the so-called ozone non-attainment region. EPA has given guidance to the states on this, and the states are now putting the first set of rules in place. Those rules allow fuel switching and efficiency. We believe that this will result in substantial CO_2 reductions and increases in efficiency.

The second issue was mentioned by the Secretary. That is global climate change issues. We are approaching it by fostering technology and allowing market forces to work through voluntary programs like the star program. This is the way to go. They have been successful programs and we support them. AGA has our laboratory signed up to the Green Lights program. We like to conserve electricity.

Third area – vehicles. That's where the oil is used. If you have an oil or energy security problem, you have to go at the transportation sector. Arithmetic tells you that. We've had a discussion of alternative fuels. We think what's going on in the initiative through the clean cities program that was announced last week is great stuff. It's going to make a real difference. We have 30,000 natural gas vehicles out there, and 800 stations: that number is increasing almost a station a day. Infrastructure is coming into place, you've got core infrastructure in a number of cities. The Clean Cities program will result in core infrastructure in more cities. Once you get that demand pull, that's going to result in OEMs making equipment, and Hazel O'Leary referenced what we saw from the OEMs, it will make a difference. This is not a market that develops overnight. You've got nearly 200 million vehicles out there. When you start with 30,000, you have a tough row to hoe.

The final issue, and I will support the secretary on this, is R&D funding. The country has maintained its leadership, the energy industries in this country have to maintain their leadership, through cutting-edge technology, and the DOE needs to be part of a partnership with the private sector, and we support this. Overall, we're here to support efficiency out of self-interest, but I think efficiency people will take that. Thank you.

Wamsted: Thank you, Mike. Our second speaker also comes from the federal sector. He will be able to clue us in on the second component of the national action plan, which will be run and directed by the Environmental Protection Agency. John Hoffman is currently director of the EPA's Global Climate Change Division. He has been with the agency for twelve years, played a key role in the development of the Montreal Protocol in the 1980s, and has now turned his sights on global climate change. Please join me in welcoming John Hoffman.

John Hoffman: I'm going to talk about three things. Why should we believe these programs are going to work? How can you profit from these programs? And what can we do together to make all this work?

I just want to quickly tell you what programs the EPA sponsors. Some of you may not be aware of them. We plan to make them looked upon as household brands, like Nabisco. In the commercial/industrial buildings area, we have Green Lights, which is a program to get more energy-efficient lighting installed. We have the Energy Star Buildings program, which is for the HVAC (heating, ventilating, and air conditioning) equipment. We have Energy Star computers, printers, monitors, and other office equipment areas. In residential, we had the Golden Carrot refrigerator program, where we worked with the utilities to have demand-pull to get new technologies to market. There will be dozens more programs like that, and additionally, programs to improve the delivery of infrastructure and financing to residential areas.

In supply productivity and efficiency, we're working on better transformers, landfill generation, and the use of generation from animal and agricultural wastes. I'd like to also state that when I talk about our programs, we really view these programs as partnerships with the Department of Energy. We've done a reasonable amount of work together in the past, but we're absolutely committed through this action plan to really make our efforts be coordinated and to support each other, and they have a number of sister programs and research efforts that will feed into what we are doing.

But despite this, there's really a question that you have to ask yourself. "Why believe that these are going to make any difference?" We've had energy-efficient products around for a long time, but the light bulbs haven't jumped into the sockets. We've had opportunities to make pretty good rates of return, but people haven't taken them. So what's going to be different here? We've just announced some programs, we've got some things happening, but why believe that this is going to change some people's behavior?

Well, for one thing, these programs are very different from those of the past. The Green Lights program and others like it are customer-oriented. We go out and we sign contracts with companies, governments, and nonprofit institutions in which they promise to do something within five years. They promise to upgrade their facilities. In developing that program, we spent a lot of time understanding the obstacles inherent in organizations. The barriers facing a facility manager. The barrier that the chief financial officer has to be able to finance these changes, or understand why they will be profitable. The barriers people have in understanding what to do and how to get it. The barriers involved in moving their organizations forward to really support these systems and actually have things happen. We developed a marketing program, an implementation support program, that meets the needs of the people we view as our customers, our partners. And so, there's a real difference between going out and preaching about energy efficiency, and going out and signing contracts with people in which they commit to do it.

Our programs are all sales programs. The people on my staff went on a retreat a couple of weeks ago, and we talked about how we are sales agents. What does a sales agent do for the customer? We really view them that way. When they sign we view them as a customer, they get an account representative. We have a standard, for example, where we return a phone call within three hours; we try and run like a business and give our customers real service. They tell us they appreciate it, and there's a real difference. We try to provide the motivation, know-how, financing, and organization to really make things happen.

We've got a track record: there are over 1,100 institutions that have signed a contract with us. They have 4 billion square feet of space for Green Lights, that's 120 buildings. Most of the orders that are going to come out of that haven't hit the marketplace. We have a track record of signatures, and if you look at our implementation monitoring, we spent about $150,000 per year just taking in the data of what people are doing. We see a curve of product orders that's exponentially rising. Within two years the market should double in the lighting sector. We have made similar kinds of agreements with computer companies: with 100% of the component companies – Intel and the other chip manufacturers – and 80% of the computer manufacturers and computer printer companies signed up to produce a computer that will go to sleep. When it's fully installed across the country, when the stock is turned over, it will save the equivalent amount of CO_2 of five million cars. These are signed agreements where people have to deliver the product, and if you look in the business section of the *Washington Post* on Monday and you start to page through those ads for computers, you'll begin to see that MicroCenter, out in Fairfax, is advertising Energy Star computers. You've seen that picture, and it's an attribute that their customers want and that they're going to make money from.

We've got 30 utilities that joined in this first Golden Carrot. Whirlpool won that competition and it was highly featured when the President and Vice President announced the action plan. Those guys went back to Benton Harbor ready to roll out their marketing programs and to invest more money in energy efficiency. They see significant profits.

We're going to help expand those programs, not just here, but abroad. We're going to be converting a factory in China to make a super-efficient refrigerator, so real things are happening. There's a track record here, and these programs are working. We are on a tremendous learning curve, and we've really made a lot of mistakes. Mistakes are part of the path to success in this organization, and we encourage people to do things, to get out there and try things.

We're understanding some of the implementation problems that some of our Green Lights partners are having with these technologies. We try to provide them with the services that they need to implement these strategies so that the customer is happy. We're getting better and better. I'm able to put a sales agent out on the street a lot faster than I was three years ago. We know how to do it. We know how to train people in sales. Their performance is based on how many customers they brought in: they don't have general agreements like, "do a good job with a good piece of analysis that the director thinks is wonderful." The contracts say "250 customers." Our resources have been relatively small. We've had about $8 million per year. Next year the agency is going to commit 300% more, and with the President's additional commitment in 1995, we're going to be 1000% more. I guarantee you that this is going to add a tremendous amount of momentum to our sales effort.

How can you profit? Well, you can look at us as a strategic marketing organization. We're really getting out there and getting commitments. You're going to deliver the equipment and service sales. Our goal is to make it easier for you to sell things: to get greater penetration and to reduce your cost of selling. That's really our goal. I think the benefits of new technology are going to be greater than ever before, because we are going to make sure that you can deliver that product to the market in a way that you've never been able to deliver it before.

Your customers are going to be demanding more and more. 20% reductions in buildings is not going to be good enough. We're going to be looking for 40, 50 and 60%. You're going to have to look at how the market will change. The character of who is making the decisions is going to change an awful lot. What I really would emphasize is for you to focus on what you do best, and invest to do it better.

What can we do together? We really want more help in understanding the barriers. If you're out selling a product, we'd love to hear from the salesmen and the people that are developing these products, what the barriers are. If you've got a good product and you're not selling it, let us know about it, because we want to understand your barriers and your difficulties. That's the way we are going to serve our mutual customers better. If your company isn't a member of our programs, join. There's nothing better than to practice what you preach. I've been fascinated by the number of energy efficiency companies who are in energy-inefficient facilities and seem not to be concerned about it. Your buildings should be showcases. Become allies. We have allied programs for almost everything we do.

Strategic marketing is really important. We had our first supplement in *Fortune* this month. Not the *Building Manager's Magazine,* which is a great magazine that goes to facility managers and helps bring up their consciousness, but we had it in *Fortune.* Because we're focusing on chief officers, chief financial officers, business managers, to view this as a business decision that is key to their competitiveness and profitability. I know this is going to be difficult for some people, because they've seen their traditional market channel as being the facility manager or the plant manager. Real customers are going to be people in the company who want to downsize it, who want to outsource, who want to make more money and be lean, mean, and profitable.

Wamsted: Our third speaker this morning is William Nitze, the President of The Alliance to Save Energy (ASE), and I can imagine that Bill is a happy man today. Over the past couple of days a lot of the things that he's been arguing for over the years have come true with the national action plan. He came to the Alliance to Save Energy from the Department of State where he helped do some of the early negotiations in what became the Rio treaty, serving from 1987 to 1990 as the deputy to the assistant secretary of state. He worked for 14 years for the Mobil Oil company before that. Please join me in welcoming Bill Nitze.

Bill Nitze: Thanks very much, Dennis. I can't really say enough how happy I am being called the belle at the ball after having been the wallflower through the 1980s. As the President of an organization called the Alliance to Save Energy, the last few days have really been a wonderful time. President Clinton has set a very ambitious, but I believe, achievable target for reducing greenhouse gas emissions in the United States. As Cathy Zoi mentioned to you earlier this morning, the Climate Change Action Plan is based almost entirely upon voluntary as opposed to mandatory actions. Indeed, the approach taken is very consistent with the approach that the Alliance has taken since its founding in 1977. From the very beginning, we have sought to build coalitions among different interests in government, in the business sector, in the environmental community, among consumer and community groups, even in the union community, to support specific policies and programs to advance energy efficiency. Some of this work has been directed at federal or state legislation or regulations. The Alliance was very active in helping to draft the provisions in Title One of the Energy Policy Act of 1992, for example.

Much of our work has simply been to try and make energy efficiency happen at the grassroots, through things like better home weatherization, better deployment of efficient lighting systems, and better organization and management of utility demand-side management programs.

Now, as we look at the specific actions to be taken by the administration and its partners under the action plan, we note that approximately one-half of the 109 million tons of greenhouse gas emissions to be eliminated will come from reduced energy demand. Perhaps more importantly, more than three-quarters of the emissions reductions from the energy sector come from efficiency improvements.

Now this raises the question as to whether indeed voluntary actions can achieve this level of reduction. There is a question about the technical potential to achieve this level of efficiency improvements. I think many of you will have read the famous article in *Scientific American* a few years ago by Amory Lovins and Clark Gellings, in which they identified opportunities to reduce U.S. electricity demand by between 25% and 75%, at a cost per kilowatt lower than the cost of building additional electricity supply capacity. Amory's number is perhaps a little high, but at any rate, a 25% reduction at 4 cents per kilowatt, or perhaps a little bit more, is very reasonable. Similarly, we've had exhaustive studies funded by DOE that have demonstrated that we could increase the average fuel efficiency of light vehicles over the next decade to something like 33-34 miles per gallon, without major additional retooling expenses or other incremental costs on the automobile industry. Finally, we know that there is a whole host of supply-side technologies which will improve the efficiency of energy production. These range from more efficient gas turbines, to new natural gas technologies at the end use, such as gas cooling, to new electric technologies such as electrocoating, to technologies that have been with us for a long time, such as district heating, to better design and redundancy in combined cycle systems, better controls, and so forth. The challenge is overcoming the tremendous inertia in the system against utilizing this potential.

Let me give you an example of this inertia. When the Alliance moved into its new offices at 1425 K Street, we were able to install efficient lighting systems with occupancy sensors and other modern electronic devices to cut our lighting bills. When we tried to put new solenoid valves in our old heating and air conditioning units, we were basically told by the building engineer, "That's fine, you put them in and then when your lease is up five years from now, you take them out and put the old ones back." That's something that as a small, non-profit organization, we simply could not commit to. That example illustrates the real nature of the challenge. Every commercial building agent here, and every commercial building owner and operator in this country, has to be educated in some detail about the economic potential of energy efficiency in their buildings, and about the ways of organizing an incentive structure to make those investments cost-effective and affordable.

Basically, to overcome this inertia on a broader scale, there simply is no substitute for hands-on engagement through new partnerships. Government agency by government agency at the federal level, state by state, city by city, company by company, community by community, and ultimately, household by household.

We at the Alliance have done a number of studies with major utilities like Niagara Mohawk or Pennsylvania Electric on what really happens in the field when you get low-income homeowners to improve the efficiency of their homes. What we've found is, it makes a tremendous difference whether you simply send people information, whether you pay them one visit and really go through what it takes to wrap a hot water heater or replace a shower head or put some weather-stripping in or some insulation into their attic, or two visits or three visits. The curve goes sharply up. In three visits, we found that in certain rural communities you can get 16% sustainable energy efficiency improvements with paybacks of certainly no more than a year and a half, but it is the hands-on engagement that is critical.

Finally, we've got to extend this out into the next generation. We're working with Washington Gas and with PEPCO to try and reach homeowners, and in part their children, through the school system. This is very difficult work. Children have many other things competing for their attention. Parents have many other things competing for their attention, particularly in lower-income neighborhoods. To overcome this barrier, we plan to greatly expand our work at schools in basic energy education.

Now, what specifically is the Alliance doing to help implement the plan that was announced yesterday? Let me just run through the major sectors and list some areas that we're giving priority to.

Probably the most important area for the Alliance, and the one in which we've traditionally been strong, is the residential sector. There, we are focusing on two important areas. First is to promote home energy lighting systems and energy-efficient mortgages. This has been a challenge that has been with us for fifteen years. There have been a huge number of interests at the table, and the electric and gas industries have had some difficulty working together on what constitutes an energy-efficient home, for reasons that I'm sure many of you understand. Primary lenders have not been interested in going through a lot of additional paperwork to finance energy efficiency improvements. Developers have been extremely adverse to anything that increases the first cost of a new home. But somehow, we're going to work through this challenge, and our goal is to make sure that a sizable percentage of homes, at least 10%, hopefully more like 20 or 25 percent, have energy-efficient mortgages and meet minimum efficiency standards.

The second area is to make sure that all 50 states adopt codes that are at least up to the model energy code standard as soon as possible. We recently received a grant to work in Ohio, and we will be working in other states that are in the lead on this, including Pennsylvania, but it's going to take a major push. There is simply no way that the Department of Energy and the administration can reach their goals under the Climate Change Action Plan unless we have a substantial upgrading of building codes across the country, and unless those codes are enforced. There is a major problem when it comes to enforcing building codes in this country. Additional resources are going to have to be put into enforcement at the state and local level.

Let me now turn to a second priority, which is the commercial sector. We are working with DOE and EPA to implement many of the voluntary green programs that you've already heard about this morning. One of our key foci is the Green Buildings program, and specifically retrofits to the huge existing stock of commercial buildings. We just have to create a set of incentives, lease terms, financing arrangements, and management arrangements that overcome the kind of problem that we had at 1425 K street. We simply can't have building engineers across this country saying "You can put it in, but you've got to rip it out in five years." One of the first steps we've taken to move beyond that is we've worked with lawyers for major developers to develop new lease terms to give landlords and tenants an incentive to actually make cost-effective efficiency retrofits in their buildings.

In the public sector, the revolving energy efficiency fund for public buildings, which is an important element in the President's plan, really builds upon the Alliance's work on federal energy management, which was conducted by my colleague Mark Hopkins. So it is with some pride and enthusiasm that we look forward to working with the administration in making that revolving fund concept a success.

Thirdly, the industrial sector. Here I'll briefly mention that we are a charter member of the administration's new motor challenge program, and look forward to working with them for more rapid deployment of efficient electric motors, which are a major energy efficiency opportunity throughout the American economy. Secondly, we are seeking to document success stories of energy-efficient process changes for small- and medium-sized companies in different industries and different parts of the country. One of the problems with promoting energy efficiency in the industrial sector is that there isn't enough readily available information about what has worked. Frankly, there are confidentiality problems in industry about making some of this information available. I remember when I worked for Mobil, I couldn't get the company to disclose some very simple techniques for changing over drilling muds and therefore reducing the need to use additives, because they were afraid that Shell and Amoco and other companies would garner a commercial advantage from this small improvement. To overcome that will not be an easy problem, but we have to try.

Finally, in transportation, we are working with DOE Region 3 on trying to get alternative energy technologies for vehicles into the marketplace at selected cities, as part of the Clean Cities program.

In closing, can it be done? Yes, it can be done. What it will take is a whole series of new partnerships, led by a new coalition of interests that includes the energy efficiency industry and groups dedicated to energy efficiency, natural gas, renewable energy and other clean energy technologies, community and consumer groups dedicated to improving quality of life, and governments at all levels. We've made some important progress in creating those coalitions, and we at the Alliance look forward to working with all of you in furthering that progress.

Wamsted: Our final speaker is from the state of Maryland, and will be talking to us about the efforts of the state to actually make work what Bill referred to just a moment ago. Our speaker is Gerald Thorpe, director of the Maryland Energy Administration, which in Maryland is a cabinet-level position that reports directly to the governor. He's got some good background in this area and recently implemented a statewide strategic energy efficiency plan that I hope he will discuss today.

Gerald Thorpe: I am here to give you a very brief account of the unique things that have happened in Maryland over the last two years.

States share similarities, as well as differences. They often have characteristics allowing them to be more flexible and innovative than agencies of the federal government. The experiences involved in implementing public policy in one state can often be successfully applied in other states, as well as at the federal level. Just as often, however, such experiences prove not to be transferable. Attempts to innovate, however, even when they do not work, can enrich and inform the policy process, and that's one of the reasons, I think, for my being here. I hope that what I have to say today will have some relevance to you.

I was asked to talk about one state's experience; what contributed to the rapid surfacing of the strong and independent energy presence in the state of Maryland. I would like to describe the developmental stages the energy efficiency function went through in our state to give you a sense of the magnitude of change that took place over a two-year period from early 1990 until today.

First, there were some major triggers. Maryland has a governor, William Donald Schaefer, whom many of you do not know. He is controversial. He speaks his mind; he tells people what he really thinks, and he acts quickly and decisively. He is a governor, I believe, who has a great deal of vision and is open to divergent thinking. He is not afraid to try, and he is not afraid to fail. The second trigger was a state fiscal crisis than began in 1990. A third was a very intrusive, not necessarily bad, but nonetheless intrusive set of federal legislative and regulatory acts that began to impact the states. Most notable among these for our discussion here is the Clean Air Act. The final trigger was pulled when the government attended the National Governor's Association meeting in the autumn of 1990. There he listened to Carl Sagan and other speakers talk about the environment and its relationship to energy efficiency.

The governor came back from that conference determined to do something about energy efficiency in our state. He responded quite publicly to initial skepticism from some of his cabinet members, "I know that some of you don't take these issues seriously, but I do and you should start." As mentioned previously, the governor is clear about his expectations.

Once the mandate for a focused effort on energy efficiency was made explicit, we went through several identifiable stages. The governor issued an executive order and he named the state's first energy czar, who happened to be me. He vested this new cabinet level position with decisive executive authority, not only to work with state agencies, but to command their energy resources in order to launch a coordinated energy efficiency program in Maryland. Perhaps most importantly, the governor constantly reinforced his desire that all state government employees, from top to bottom, make energy efficiency a top priority.

The governor then established a sub-cabinet group for energy management, which included the seven major state agencies of state government most directly involved with issues of energy and energy management in Maryland. The governor also established a 20-person citizen's task force, which included the full range of stakeholders in the energy arena. The task force began its work with a summit meeting of over 500 citizens to solicit ideas for major energy policy initiatives.

Legislation was introduced at this time to establish the Maryland Energy Administration as a statutory unit of government, reporting directly to the governor as an independent agency. The bill was one of four reorganization bills introduced into the General Assembly during the 1991 session, and the only one that passed. The legislation survived by only three votes in the Senate. Not only did it barely pass, but the proposed budget was reduced significantly during this period of fiscal austerity.

The message was clear. If the Energy Administration were to survive, it had to build strong credibility with the General Assembly and the general public. We took the period between the end of the legislative session in April of 1991 and the beginning of the 1992 legislative session beginning in January to work on this. Two of the things we did, I think, are most noteworthy. We resized the agency by reducing our permanent complement in favor of hiring technical experts on contract. This increased our flexibility and our ability to shift our skill mix to take advantage of new opportunities. We also adopted a matrix management structure designed to make the Energy Administration capable of responding with an entrepreneurial style and making the maximum use of a small professional staff.

We set out to deliver a simple message to the legislature, "We can save the state money." A small investment of public funds will equal a substantial return on that investment. We didn't get fancy, we focused on the bottom line – money – a substantial return for a modest investment. We had a hook – the federal EPA Green Lights program. This program proved to be not only an excellent program in its own right, but it provided us with an image rich vehicle for convincing decision-makers in both the public and private sector that energy efficiency can return substantial profits from modest investments.

So we used the Green Lights program. We built a database to track energy use in all of our agencies. We then went back to the legislature with proof that we saved the state over $2.5 million in our first six months of operation. This relatively modest beginning to a very positive legislative response. In a period of just two legislative sessions, the following legislation passed: a joint resolution committing the legislature to strong support for energy policy designed to enhance environmental quality, promote economic growth, and reduce reliance upon foreign energy supply; a bill that requires state agencies to reduce their energy use by 15% by 1996 and 25% by the year 2001; and a demand-side management law that requires our public utilities to provide demand-side management rebate programs approved by the Public Service

Commission. The demand-side management program has been infinitely helpful to us, because the utilities have become partners in helping pay for capital improvements for energy efficiency.

The Maryland General Assembly also enacted life-cycle energy efficiency requirements for major equipment purchases and design standards for new construction. In addition, two tax-incentive bills were signed into law to encourage the purchase of alternative-fueled vehicles and refueling equipment. Along with this, the governor signed an executive order directing state agencies to convert the state fleet to alternative fuel, beginning this year. Our target is for 25% of all new purchases to be alternative-fueled vehicles by 1998.

Perhaps above all, we were able to amend our procurement law to allow performance contracting. These changes in procurement law allow us to hire one expert to design, install, finance, maintain, and manage our buildings in exchange for a portion of the resulting energy savings. This represents a major change in procurement law, yet only one vote was cast against it in the General Assembly.

Over those two legislative sessions, the Maryland Energy Administration was also given several new financing tools by the Governor and the legislature. One is a state building revolving loan fund that allows the Energy Administration to loan money to agencies for energy improvements. We loan them the money and they pay us back out of energy savings. Another program called the Community Energy Loan Program, allows us to loan money to nonprofit organizations, hospitals, and local governments under the same principle. We loan them the money, we do the engineering analysis, they pay us back for the energy improvements out of avoided costs. Both of these programs are capitalized at over $2 million. They are fully subscribed and working well. The Energy Administration also received bonding authority from the state, which will allow us to make significant amounts of money available to state agencies for capital improvements designed to enhance savings through energy efficiency.

There are continuing challenges to be faced, not the least of which are significant bureaucratic resistance, turf problems, skepticism, and plain old inertia. Even when the governor says he wants something done, or the President orders it, or the legislature or Congress mandates it, it is not necessarily going to get done. So you have to constantly work on overcoming resistance and inertia. There is a persistent public skepticism. There is a lack of sustained interest in energy efficiency. There is also the short attention span of political actors. They come and they go. Every new administration, whether it's at the federal, state level, or the local level, starts with "let's reorganize." And so, we are on top today, compared to where other states are, but we could be back on the bottom again tomorrow.

Let me leave you with what I consider to be the key elements in process of establishing a viable energy efficiency program. First, the time is right for states. Things are beginning to fall into place. Devolution and decentralization are in ascendancy. We have a lot of allies and we have the Clean Air Act, which is now driving energy efficiency initiatives.

The key element for a successful energy efficiency initiative is, however, strong executive support. If the governor is not actively behind a state program and the person in charge of the energy function does not have direct access to him, you will likely not succeed.

Secondly, you need to prove to the legislature that you can save money – that money invested provides a significant return on the investment. If you can document that energy efficiency programs save public and private money for reinvestment, you'll be given a freer hand to achieve energy efficiency and improve environment quality.

Last, but not least, you have to capture citizen imagination and support by being entrepreneurial and different. You have to be able to prove you can do more with less, be opportunistic in the good sense of the word, and demonstrate that you really are worth the tax dollars spent to support you. Mobilizing your natural allies and putting them to work in order to achieve your goals makes everything easier.

THE WISCONSIN ENERGY INITIATIVE

John Glenn, U.S. Senator, State of Ohio.
Tommy G. Thompson, Governor, State of Wisconsin.

John Glenn: The State of Wisconsin, through the efforts of Governor Tommy Thompson, has already paved a path we can follow to use energy more wisely, and cheaply, to save money. Governor Thompson took the initiative and created the Wisconsin Energy Initiative. It sends out a comprehensive approach to energy management and conservation, and includes among other things, an alternative fuel program. They have 300 state vehicles, and they are testing eight alternative fuels. They have grants to local schools to purchase energy conservation systems, and they have weatherization services provided by a network of twenty-seven local organizations to improve the energy efficiency in over 9,000 low-income homes. The initiative also establishes goals to improve energy efficiency in state and university buildings: it calls for a reduction of 20% in the rate of electricity consumption by the year 2000 in state buildings, and a 15% reduction in the overall rate of energy use by state agencies. New energy efficiency standards will be developed for all future state building construction. State heating plants will see a 50% increase in cogeneration, and 100% recycling of ash. Finally, the state is conducting energy audits of all of its buildings, lighting and conservation improvements are being identified, and there is a $50 million pot of money from state matching utility funds to carry out these retrofits. This is a really active program.

The Wisconsin Energy Initiative has already begun to pay dividends. Energy efficiency is up, taxpayer monies are being saved, jobs have been created, and the state's environmental quality has been improved. Any program that can address such things as energy efficiency, taxpayer money, jobs, and environmental quality at the same time, is a superb program. The Johnson Controls / United States Energy Association Energy Award is a fitting tribute to a man who has demonstrated leadership, not just by word in talking about the energy problem, but by actions. One who has shown us that energy conservation goals can be met. They can be met cost-effectively, and without any sacrifice of comfort, and I know that he has spread this gospel far and wide to all councils of government within his state. I'm glad to see him being honored this evening so this message gets far beyond just his state.

I think that it is very fitting that I, as a Democratic senator, present this honor to one of our more noted Republican governors. And there is a good reason for that, for we both understand, that energy conservation is not a partisan matter. It's something we have to have for this whole nation, it is a good government issue, and so it is my honor, and with the most sincere congratulations, and in the spirit of bipartisanship that I am very pleased to present the 1993 Energy Leadership Award to Governor Tommy Thompson.

Tommy Thompson: It's quite an honor for anyone to be recognized for what they do for conservation, for the savings of energy and for the environment. On April 23rd of last year, I announced the *Wisconsin Energy Initiative,* which was, as the senator indicated, a $50 million investment in energy conservation as well as environmental protection for the State of Wisconsin. On that very same day, I was proud to enter a partnership, and I believe these partnerships that we start between the public and the private sector, are what makes Wisconsin such a wonderful state. We invited Johnson Controls to become our partner, and we gave them a contract to supervise our Energy Initiative, and what an outstanding job they've done. It makes sense, it makes common sense, to be able to enter into public-private partnerships that work. Johnson Controls, of course, is a world leader in energy management systems and, of course, they've done a great job with our Energy Initiative. It's because of this strong partnership with companies like Johnson Controls that Wisconsin, I believe, has been able to lead the way in so many different areas.

We have a proud tradition of protecting our environment in Wisconsin. We have some of the strongest environmental laws in the books. Our clean air and clean water laws served as models for the nation. We passed our clean air act five years before the Federal Clean Air Act, and our reduction of utility emissions has already surpassed the quotas for 1993. In fact, we are already meeting the quotas for the year 2000. We set up a regional fund between several states, the Great Lakes Protection Fund, to clean up and improve the Great Lakes, funded by $100 million of state monies from the respective Great Lakes states. It was the first model of that type in the country. We are preserving vast areas of Wisconsin's wilderness for future generations.

All of this and, at the same time, our economy is thriving. We've been able to cut all taxes and, of course, if you're looking to expand your business, Wisconsin is the only state that has a 60% exemption in capital gains, no inheritance taxes, three cuts in unemployment compensation, the best workman's compensation laws, and a lowered income tax – and we still have more revenue than we had before. Wisconsin is proof that you can protect the environment and create jobs at the same time; we did it in Wisconsin and we can do it all over America. The *Wisconsin Energy Initiative* is just one more step along the way, and we're seeing results.

In the first year and a half of the initiative, energy conservation on University of Wisconsin campuses has already totaled $550,000 a year. The state has received more than $1 million dollars in rebates from our utilities, we expect an additional $800,000 next month, and another $1.4 million in rebates has already been identified. The state is using recyclable paper pellets and wood chips in our heating plants, and the paper pellet project has generated four new businesses that sell pellets, and with whom we contract for our state's heat. Wisconsin was the first state to receive alternative-fuel cars into our fleet from General Motors. And working with the Council of Great Lakes Governors, we've formed a consortium for state governments to buy recycled copy paper. A Wisconsin paper firm is now producing recycled copy paper for our state, as well as for Indiana and Michigan. This initiative saves trees, energy, and landfill space. It also saves the State of Wisconsin about $25,000 each year. Energy conservation is working in Wisconsin, because it simply makes common sense.

ENERGY FOR TOMORROW'S WORLD

Gerhard Ott, Chairman of the Executive Assembly, World Energy Council.

Moderator: Our keynote speaker has the distinction of being able to speak for over 100 developing and industrialized countries on energy matters. Doctor Gerhard Ott was elected Chairman of the Executive Assembly of the World Energy Council in 1988, and was unanimously re-elected in 1991. A non-governmental organization founded in 1924, the World Energy Council's objective is to promote the sustainable supply and peaceful use of energy for the greatest benefit of all people. Dr. Ott will present a report of inordinate significance that will have far-reaching consequences. The report by the World Energy Council Commission is entitled, "Energy for Tomorrow's World: The Realities, the Real Options and the Agenda for Achievement." It is a multi-year, $5 million comprehensive global study that provides pragmatic options for change. These ideas must be considered if we are to achieve successful and sustainable development in the world.

In some ways, the subtitle of this report, "The Realities, the Real Options and the Agenda for Achievement," provides a brief synopsis of Gerhard Ott's life. Dr. Ott has faced the realities of the business world in his role as the Managing Director of the Gas Union (GMbH) in Frankfort. He has examined life's real options throughout his education at the University of Munich and the University of Michigan Law School. And, finally, through his achievements as Executive Director of the 11th World Energy Conference in Munich and his leadership at the World Energy Council, he is helping to set and define the agenda for a sound energy future. Ladies and gentlemen, please join me in welcoming our keynote speaker for the 4th Annual Johnson Controls/U.S. Energy Association Energy Conservation Forum, Dr. Gerhard Ott.

Gerhard Ott: Thank you for this very warm introduction. Ladies and gentlemen, I am afraid I have rather bad news for you. I come from Essen, Germany, Continental Europe, and I must admit I am not familiar at all with the fine art of garnishing luncheon addresses with nice jokes like our British friends do so well. My speech, I'm afraid, should be taken seriously, and I hope you really have enough patience.

By now I think you know the book, and I am grateful, Paul, that you mentioned the subtitle, which is very important for me. It suggests doing something of substance, close to the earth, not something up in the clouds. As a study, which will be available soon in ten different languages, I think this is an achievement of its own.

Let me ask what directly the World Energy Council study has to do with the topic you have chosen for your forum, "Energy Efficiency, the New Agenda." I think very much, indeed, because this study really has made a first attempt to bridge the sometimes very deep trenches separating those who represent energy from those who see their main commitment to defend the environment. In doing so, the World Energy Council has identified energy efficiency as the certain common ground between those two parties, and I'm confident, therefore, that presenting this study will not sidetrack from your issue, but contribute to it.

Energy for Tomorrow's World is a study commission based on grassroots work in all regions of the world, coordinated by a central management unit which was based in London, and advised and guided by a board of some 50 eminent persons, some of whom I see here. Presenting a new energy study very often provokes a very basic and simple question: "Why yet another study in a world which is already flooded with studies, government documentation, reports of this sort, etc.?" Let me give you two main reasons why we decided to take on this very demanding, ambitious, and also very costly task.

For one, the World Energy Council's last comprehensive report was issued back in 1977. It was sparked by the experiences of the first oil crisis. Obviously, the world of energy has changed considerably since then, making a new assessment mandatory. Secondly, and maybe more important, in response to the publication in 1987 of the well-known United Nations World Commission's Report on Environment and Development, "Our Common Future," the World Energy Council wanted to ensure with this new study that energy, per se, was properly represented.

While that study, the Frontland Report, raised the whole question of the earth's sustainable development, it did not, in our view, give proper consideration to the role energy might and should play in this context. Neither, by the way, did the two main documents produced from the Earth Summit in Rio De Janeiro in 1992. In this context, I think the new WEC study has an important function to fulfill, to complement the Frontland Report through highlighting the role of energy in a sustainable future for our rapidly increasing world population, and to correct some of the too high expectations created at the Rio summit.

Here are a couple of examples. Instead of offering the rather unrealistic prospect of some 125 billion U.S. dollars annually to be transferred from the developed world to the developing world, our study emphasizes the need to create more local and domestic finance. And instead of merely pleading for a generous and free-of-charge transfer of technology from the industrialized to the developing world, this study puts emphasis on the preconditions for any such transfer to be established beforehand in these countries, such as setting up a reliable legal framework, protection of intellectual property, etc. Such a realistic, down-to-earth approach may be sobering to some who prefer to indulge in high-flung ideas about an altogether altruistic world. It is my conviction, however, that only a practical approach will achieve practical results. At the same time, I'm convinced that such a practical approach carries its particular weight coming as a consensus of the World Energy Council, an organization with some 100-member countries, and more than half of them representing the developing world.

So much about the background of the new study. It is not, let me underline again, just another computer-driven demand/supply exercise by extrapolation. Rather, it is a study undertaken by energy operatives, undertaken by an independent, non-governmental, multi-organization energy corporation. And it examines regional priorities in use before synthesizing them into a global report.

Let me try to coin the main findings into seven simple principal messages. **Message one would be: sustainable development calls for more, not less, energy.** The population of this globe today is 6.5 billion people, and it will continue to increase into the 21st century, up to over 8 billion people in the year 2020, and possibly up to 10 or 11 billion in the year 2050. This is what we have to face, and if we are to avoid mass poverty, fighting for access to resources, tensions, catastrophes, political disruptions, then we simply need more energy, and above all we need more energy for the developing world. Only an adequate supply of this energy can guarantee the minimum of economic growth, which is indispensable for feeding, housing, and educating this growing world population, but also, not to forget, paying the price for better protection of the environment. Let us be very clear about one thing which has been shown very clearly by the report: Energy and environmental problems are shifting more and more to the countries of the Third World. Around the year 2020, the developing countries will account for more than 80% of the world's population. They will claim at least 85% of the global energy consumption increase anticipated for the next decades, and they will be responsible for some 70% of global SO_2 and NO_X emissions, and more than 50% of global CO_2 emissions. Obviously, therefore, sustainable solutions for developing countries cannot be transplanted from outside, they must come from within.

Message two: The global energy mix will change only gradually and only over time. The various forms of energy as additives, rather than alternatives. And to take away one illusion which is rather common nowadays, renewable energy sources will, up to the year 2020 and probably for some time beyond, at best, supply a modestly growing part of the additional demand for energy and, under no realistic circumstances, will be able to substitute for oil, coal, gas, or nuclear. I say this despite widespread belief to the contrary, and even with strong assistance from government policies. All forms of energy, I repeat, must be seen as additives not alternatives. Later, after regular decrease of oil and gas (which might start around the year 2050), it will be coal and nuclear that offer the best options for expansion. Before then, it's a very far future when completely new energy sources and energy systems might lead us into a completely new era of energy production and use.

Message three: CO_2 emissions are bound to increase far into the next century. This is not exactly the story we heard this morning. I'm talking from the global viewpoint. An increase in energy demand globally by 50% or more over the next 20 to 25 years and the continuing dominance of fossil fuels within the global energy mix, are the two main factors that will lead to further increases of CO_2 emissions, even with extremely ambitious assumptions as to future improvements of energy efficiency. This is the report's conclusion, even assuming reducing energy intensity in most developing countries by over one-third and in industrialized countries by one-half. There is no realistic possibility to contain or to reduce greenhouse gas emissions and atmospheric concentrations on a global level during the decades to come. Such a development is conceivable only toward the end of the *next* century, as the study's long-term projections indicate. It remains to be seen what effect will come from the UN Framework Convention on Climate Change, signed now by some 160 countries, but probably unrealistic in its expectations. And it remains to be seen what will come out of the action plan just announced in this country and a similar plan just announced in my country.

But to avoid any misunderstanding, ladies and gentlemen, the World Energy Council certainly is not against such high-aiming and certainly noble objectives. We do, however, prefer a realistic assessment to wishful thinking, and instead of waiting for miracles to happen, we believe in a practical response against imperative impacts from possible global climate change. And such a response, in our view, would have to comprise firstly – improvement of the scientific basis of the climate discussion; secondly – determined efforts to increase energy efficiency; and thirdly – timely and prudent measures of adaptation. This, by the way, is very much in line with the approach taken by the Inter-Governmental Panel on Climate Change for its second assessment plan for 1995.

Message four: Environmental concerns with local and regional impacts for countries have a higher priority than possible CO_2 effects. In fact, only one of the study commission's nine regions gave top priority to global environmental issues. The rest of the regions, and specifically the developing countries faced with their own priorities for using scarce resources, place the highest priority on local and regional pollution issues. And, indeed, pollution of water, land, and air; lack of sanitation, decent housing, health care and refuse disposal – these, ladies and gentlemen, are the real problems of this world. These problems are going to grow.

Let me mention just two developments which soon will add to the concerns. In the year 2020, more than half and toward the end of the next century almost three-quarters of the world population will be living in urban areas. I'm asking, are we in any way prepared for this? The answer, I'm afraid, is no. Or take transportation. In the United States, it is 1.7 persons who share a car. California, of course, is a little bit different. As for the USSR, there is one car for every 17 people, and in China close to one for 700. Are we really aware of this avalanche which will break loose with even a modest improvement of the standard of living in these countries? We have a little example in East Germany, but the global avalanche will be much, much greater.

Message five: The key for reconciling the need for more energy with the need for protecting the environment is improving energy efficiency. This message seems to be self-evident, but it is another area where aspiration and ambition very often do not match with actual achievement. The theoretical potential for improving energy efficiency, undoubtedly, is the road to success. But, it is full of potholes and hindrances, lack of awareness, lack of technology, lack of correct pricing, lack of finance, just to mention a few. Although improving energy efficiency, which means in the end using less energy, would seem to be a natural and ideal response for those countries that suffer from an inadequate supply of energy – again, mainly developing countries and countries in transition – so realities very often are different. The reasons again being lack of understanding, lack of resources, etc. Whether we like it or not, high energy efficiency is not necessarily placed at the beginning of the countries' positive economic development, as it should be, but rather at the side of such development. Or, in other words, a luxury that only rich economies can afford.

And how difficult it can be to improve energy efficiency. Let me read to you from a recent *Time* magazine article. "Many hands make light work." OK, how many government employees does it take to change a light bulb in a British National Health Service Hospital? The answer is six, and that's official, and that is no joke. The study released by the British Audit Commission reported that six people were required to get a light bulb changed. The task involved 17 different steps and took 20 minutes to organize. And it was an ordinary light bulb; just imagine how long it would have taken to change it against an energy-saving light bulb! Now this is the UK, and I'm sure in your country, where there's a new health plan and a streamlined government, you certainly will not fall into this trap.

What we need, therefore, in order to make this scheme and energy efficiency work, is not so much learned articles on theoretical potentials, but more pragmatism. For example, practical help to install step-by-step meters in millions of Russian or Ukrainian households that can substitute for the window as a sort of regulating instrument. Practical help to convince utilities say, in Africa or Latin America, is that the prices charged for electricity have to cover at least full production costs. Otherwise, a process is started that will drive those companies, sooner or later, into bankruptcy. I think this kind of practical help and advice is of better value, especially for those countries, than lectures on the great values of democracy and a market economy.

Message six: Energy will continue to be a major factor in geopolitics. I admit, considering the fascinating and, at times, merciless fight between Europe, America, and Japan for superiority in the field of high technology, one might be inclined to think that in the future it will be techpolitics only that count in creating a new world hierarchy. I feel it would be a mistake to make this assumption. However, geopolitics will continue to be a determining factor and energy, together with food, water, and other natural resources, will be an essential element within this context. This will be all the more so because, the new study has shown, the number of countries that are dependent on energy imports will increase sharply over the next decades. Fifty percent of the world's population is living in countries dependent on energy imports, and this figure will rise to over 80% by the year 2020. In the best of all worlds, this might be considered to be a welcomed opportunity for expanding world trade. Looking at the world as it is, however, one certainly cannot exclude the possibility that this growing dependency on only a very few net exporters of energy might lead to social and political tensions.

At times, ladies and gentlemen, I think there's still too much caught in our tradition. We are beginning to take a more direct and active interest in how to enhance the safety of former Soviet nuclear reactors. Realizing, admittedly a little late, that another

Chernobyl accident, in all probability, would mean the end for any further expansion of nuclear energy, which Japan needs so badly for its economic development. Also, energy-poor Japan might move faster into some of the huge oil, coal, and gas resources in the Far East and part of the former USSR, waiting for capital and know-how to be exploited. Some Western Europeans who still think of claiming these resources for their own countries, relying on the European Energy Charter might soon find, to some surprise, that these supposedly eastern parts of Europe in the meantime have become part of Western Asia.

Another more important example, perhaps, because it concerns not only Western Europe, is that all these soaring economies in the Asia/Pacific Region have a hunger for energy and, above all, an unquenchable thirst for oil. But oil is the resource that they lack most. This is true already today for Japan and India, and will become a harsh truth for China and also for the oil-rich Indonesia around the year 2000. And some of the Young Tiger countries are in a similar position. All these newly emerging economies with aggregate purchasing power will turn more and more to Arab countries of the Middle East and to the former USSR for the oil and gas they need to fuel their economy; to sources that, so far, have been considered by many as reserved for Western industries and consumers. And already today, Asian countries are paying more and more attention to Middle Eastern politics. I'm firmly convinced, therefore, that energy in the future will rank high as a factor determining worldwide politics.

Message seven: The real risks and chances lie beyond 2020. Long lead-times are characteristic for energy investments, for environmental impacts, and for population growth. Therefore, developments over the next few decades are more or less predetermined, leaving only room for marginal changes. It is only after the year 2020 and, in particular, the second half of the 21st century that population growth may come to a halt, that emission of greenhouse gases may decrease, that new energy systems based on solar, advanced nuclear superconductors, etc., may come into effect, if, and this is my point, if we start making the necessary decisions now.

If we want to avoid the grave risks undoubtedly lurking after 2020, or make use of the chances waiting there, then we have to act now. Hence, the inclusion of the study commission's epilogue, the last chapter, which highlights the many factors that may contribute to energy's potential strategic programs after 2020. It serves to emphasize that the next 30 years are, above all, ones of managing energy transition and initiating the profound changes that will be required in the global energy sector long after. The World Energy Council has ambition to work on these long-term perspectives in more detail in the forthcoming two years leading up to our next World Congress in 1995 in Tokyo.

Ladies and gentlemen, I promised at the beginning of my address that the findings of the World Energy Council's new study would be of direct relevance to the topic of this forum, "Energy Efficiency: The New Agenda." I think, indeed, the study has confirmed the objectives and intentions of your meeting – that improvement of energy efficiency is the one means that offers both environmental and economic improvement, and it merits enhanced attention by the energy community. Hopefully, your forum and our study commission will certainly lead to a better understanding of the problems involved, to a deeper insight into what is at stake.

To underline this, I know of no better way than to quote from another Conservation Commission study done in 1978, "Radical action is needed without delay. The time required for the establishment of new facilities, or significant changes in patterns of energy use, is now so great in relation to business and even political cycles, and in relation to probably doubling of demand, it will not be possible to turn off energy programs in periods of recession and expect to meet the demand, to meet the need and period of expansion. Long-term programs, and resolute pursuit are required now." And, again for me, there's not a single word to add, except let us not risk any more delays. We have already lost, in my view, too much time, which, in effect, is our scarcest resource. Thank you very much for your patience.

Questions and Answers:

Q: I understood you to say that energy efficiency and renewable resources are important, but in the longer term. That seems to be contrary to what we heard this morning. Is that a generalization and, if so, do you think the present administration is making a mistake in leading us to believe that short-term benefits are just around the corner?

Ott: No, I wouldn't really see a contradiction. Yes, I do say renewable energies and efficiency will have a growing and considerable contribution to make only over time. But, in order to achieve this, you have to start now. Absolutely, and with all the hindrances against the energy efficiency measures, it needs a lot of motivation, almost overwhelming motivation. So I would not mind at all even a little bit too high expectations right now, it just needs this kind of state of mind to get it all started. But again, this is why I make my sobering remarks, let's not expect too much too soon, because if it leads to disillusions, we can lose momentum.

Q: If you want to install meters in the former USSR, that would be a start. The question is, how would the public, residents of these villages, be taught that this is no longer a socialistic form, where it's a free hand-out. You have to pay for it. Education is required. How long do you think that would take?

Ott: Yes, it's a fair comment, and you are underlining what I have been trying to say. It is really a heavy program. I remember Russian studies very shortly before the old Soviet Union came down, saying there was an energy savings, as they called it, a potential of some 50-60%. But, I do happen to know Russia a little bit. It takes a lot of time, and you very rightly pointed out it is not only the question of installing meters, but first creating an awareness among people, and people who have to go hunting everyday for bread are not actually in the mood for taking lessons on how to meter, etc. It is possible, money is important, but, above all, it is the inner state of the mind and again time against which we have to fight, but we have to start with those little things.

I think we have made progress, because still four, five, six years ago we used to discuss energy only in terms of environment. Now we are getting more things together and we are finding that economy is another important "e," an important link. And I would add a fourth "e," education. That is really what it comes down to. How can we educate someone who is hungry? Very simple: you have to provide a certain standard of living first. This again needs energy.

Catherine Zoi from the White House Office on Environmental Policy provides details of the 1993 Clinton Administration energy efficiency plan.

Domestic and international energy concerns were highlighted in 1993 in speeches by newly inducted Secretary of Energy Hazel O'Leary and Gerhard Ott, Chairman of the Executive Assembly, World Energy Council.

Brian Atwood, Administrator, U.S. Agency for International Development, provides a spectrum of opportunities available for energy efficiency in developing countries in 1993.

Senator John Glenn (D-Ohio, right) presents the 1993 Energy Leadership Award to Governor Tommy G. Thompson (R-Wisconsin) as Johnson Controls CEO James H. Keyes looks on.

A 1993 panel discusses domestic energy efficiency issues. (L to R - John Hoffman, EPA; Dennis Wamsted, Energy Daily; Gerald Thorpe, Maryland Energy Administration; E. Morgan Massey, A.T. Massey Coal Co.; Michael German, American Gas Association; and William Nitze, The Alliance to Save Energy.

Fifth Annual

Energy Efficiency Forum 1994

Fifth Annual Forum - 1994

DELIVERING ENERGY EFFICIENCY IN THE '90s

The fifth annual forum, "Delivering Energy Efficiency in the 90s," marked a transition in the development of the USEA/JCI conference, a progression symbolized by the title of this book, *Policy Evolution.*

In 1994, the name of the event was changed from the "Energy Conservation Forum" to the "Energy Efficiency Forum," and the emphasis of the proceedings somewhat shifted away from the U.S. federal government toward a focus on emergent changes in the energy marketplace, particularly the impending deregulation of the electric power industry as a result of the Energy Policy Act of 1992.

In earlier years, there certainly had been some discussion of the economic significance of energy conservation; indeed, a Congressional Office of Technology Assessment report released during the Bush era noted that federal procurement of energy efficient products could "expand market opportunities for producers of efficient technologies, demonstrate measures useful in the private sectors and encourage more research and development by manufacturers."

But by 1994, in the wake of the Clinton administration's explicit linkage of energy policy and environmental protection policy with strategies for economic growth, the idea that energy efficiency, opportunities for business development, and "growing the economy" were fundamentally interrelated and compatible set the tone for the entire forum.

The two panel discussions, which dominated the audience's attention, both focused on questions related to the changing energy marketplace. We reprint the full discussion of the first panel, "Delivering Energy Efficiency: Is Demand-side Management In or

Out?" including the question and answer session. This panel was composed of three economists, one businessman, and one lawyer, each bringing a unique perspective to a very lively discussion.

We present excerpts from the second panel, "Promoting Energy Efficiency: Command and Control versus Market Pull," which focus on the international dimension, especially the experience of the European Union. We have also included some of publisher and long-time energy scene observer Llewellyn King's keynote remarks about the difficulty of deregulating the U. S. electric power industry.

Throughout this fifth annual forum, the big question on everyone's mind was, how will the delivery and implementation of energy efficiency fare under new market conditions that are likely to be radically different from what consumers, providers, regulators, and policy makers are used to? The answers are stimulating and pave the way for the continually changing attitudes about energy efficiency.

THE WHITE HOUSE

WASHINGTON

September 13, 1994

Greetings to everyone gathered in our nation's capital for the fifth annual Energy Efficiency Forum. I am pleased that so many leading representatives of this important field have come together to explore the latest trends in energy efficiency.

The continued health and success of our nation require us to maintain our commitment to protecting our precious natural resources. My executive order on energy efficiency, which I was pleased to sign last March, recognizes this vital mission. Mandating the use of innovative technologies throughout the federal government, this initiative will enable us to save taxpayers money, create jobs, reduce air pollution, and protect the global environment.

Energy Awareness Month provides an ideal opportunity to renew our determination to meet the energy demands of tomorrow. With the skills, knowledge, and active involvement of individuals in the public and private sectors alike, we can develop alternative energy sources, promote conservation, and work toward an environmentally and economically sustainable future. You can take great pride in your efforts to reach this crucial goal, and I commend each of you for your leadership and foresight.

Best wishes for a most successful conference.

Bill Clinton

DELIVERING EFFICIENCY: IS DEMAND-SIDE MANAGEMENT IN OR OUT?

Vinod Dar, Founder and Chair, Jefferson Gas Systems, Inc., and Chairman, Sunrise Energy Services.
George Galloway, Partner, Stohl, Reeves, Boley, Jones and Gray.
Ashok Gupta, Senior Energy Economist, Natural Resources Defense Council.
Eric Hirst, Corporate Fellow, Oak Ridge National Laboratory.
Irwin Stelzer, Director of Regulatory Policy Studies, American Enterprise Institute.

The following questions were put to the panelists by the moderator, Robert Marritz, editor and publisher of *The Electricity Journal:*

1. Are there market barriers to energy efficiency, barriers such as customer information and customer access to capital, which have often been fingered as a problem to achieving an efficient energy delivery system? and if so:
2. How do we deal with those market inefficiencies?
3. Are power rates the issue, or is it the total energy bill, which could include energy efficiency services delivered by a utility, or maybe others?
4. What structures will likely evolve in this industry and other power industries around the world as we learn from each other about what the best way might be to deliver a product which was once thought to be monopolistic but is no longer seen that way?
5. Can demand-side management (DSM) fit into a competitive model of the industry?
6. What is going to happen in all this to integrated resource planning (IRP)?

Vinod Dar: It's always a privilege to be allowed to present one's views in an open forum like this, especially when the views concern what at the heart of it seems to be a debate between economic democracy and central planning. I would actually have thought the debate had long since been resolved and we were here to decide how to implement the results, but apparently every industry must learn at its cost, at its own pace, and at the risk of bruising the egos of those in control, that at the end of day the market prevails and the best laid plans of the central planner usually crumble into obsolescence. But I guess we restart the debate today and the question that you have posed is a very useful and interesting one.

The question "Can DSM coexist with competition?" begs a second question. Demand-side management may be one of the most graceless expressions ever coined, because it reeks of central planning. I think if you labeled it for what it really is, the marketing of efficiency, or the retailing of efficiency, it then becomes a service no different than the retailing of any other service, good, or product. And if the question is, can a service be marketed or retailed better with competition than with a monopoly, the answer is absolutely yes. I would argue that not only will efficiency retailing prosper under competition and open access, but efficiency will be a bigger market under competition than it ever could be under monopoly. Cleaner power will be a bigger market under competition then it could ever be under monopoly, and the consumers of electricity will actually be called customers instead of ratepayers. So I would argue that the current system is bad for efficiency, bad for the environment, bad for consumers, and bad for the competitiveness of whichever economy chooses to stick with what I have to believe is a relic of the 1930s, and like most relics the 1930s is long due for burial.

The question is asked, "are there market barriers to DSM?" Absolutely, there are market barriers. The biggest barrier to DSM is the lack of consumer choice. The utilities and the central planners, and the philosopher-kings who sit around making policy basically have the attitude of, "Hey you, ratepayer, we have a red polyester jacket for you. We have decided that this is very good for you, and it's half-price, so why aren't you buying them?" Well, perhaps the customer is allergic to polyester, perhaps the customer dislikes red and perhaps the customer doesn't want a jacket. They didn't ask the question "Would you like to be comfortable, live better, rise in the esteem of your peers, and save money?" The answer to all of those is yes. The reason centrally planned programs fail is because they represent the bureaucrats' view of what people want, and very often that is totally at odds with what the people really want. So I think if one wants to market efficiency we should start not with the producer, the planner, or he government agency, but by going to the people who pay the bills. Many in this industry forget that there is only one group of people who actually pay the bills, they're called consumers. The rest of us simply struggle over the disbursement of those bills.

What inhibits customer choice today? The two biggest barriers to consumer choice today are monopoly control over the meter and the leveraging of the monopoly over the wires into monopoly over supply and merchant markets for electricity. Why is monopoly control over the meter bad? Competitive markets cannot work if information is bottled up and trapped inside a monopolist. As long as there is monopoly control over the meter, there is monopoly control over both form and the content of information. If one controls information, one controls every decision that it is possible to make with information. Therefore, unless the meter is liberated from the monopolist, information cannot be liberated. If information cannot be liberated, it cannot be used. If it cannot be used, the consumer cannot be empowered. Choice without information is as meaningless as voting in Albania ten years ago. How do we deal with these barriers? I think the answer is by bringing about a structure that separates all the functions, supply, transmission, distribution, and the merchant function, and makes the merchant function of electricity as unbundled as possible into services, the electric commodity itself, and the marketing of efficiency.

A question has been asked about what structures will evolve, what's important, rates versus bills, can DSM fit into a model of retail competition? We have to start, not by asking how do we induce customers to invest in efficiency, but finding out what it is they want. Based on focus group studies we have conducted in California this summer, customers tell us they want the following things: continued reliability, lower prices, customized billing, and real choice in who supplies them, and they want all these things now. I believe that if we had true retail competition in electricity many merchants would rise up offering as a package: information, electric product, and efficiency. Every customer wants something their own way, none of them want it our way, and it's only by having independent merchant competition and by liberating information that we can really give people what they want, their own way.

Let me close with this question that has been asked, and incidentally, I don't believe wholesale competition will do it, I think wholesale competition by and large is a sham if it does not lead to retail competition. Let me close with a quick response to about integrated resource planning. I believe that IRP should be scrapped at the level of the utility and at the level of the regulatory commission. It is a relic. However, there is a form of IRP that should be encouraged. It's at the level of the consumer. Every consumer who goes to the grocery store today engages in IRP. You don't have to tell them how to allocate their consumer dollar when they go to a grocery store. If freed, if liberated, the consumer is more than capable of engaging in IRP, and the collective decisions of millions of free consumers will lead to IRP as people have defined it.

George Galloway: I probably should disclose that I'm something of a ringer at a gathering such as this, and while I've had some passing involvement with energy efficiency programs, most of my time is devoted to helping people who are in the process of producing and selling electricity. So my message is a message from the world of both power markets, particularly in the western United States. I've been involved in those markets for nearly 20 years. I'm extraordinarily impressed these days with the pace of change and the dramatic extent of the change. Most notably, prices are falling, the market clearing price of electricity in the western United States is about 25% lower than it was a year ago. I think it'll go lower yet. The popular wisdom is to suggest this is a temporary phenomenon, a short-term mismatch of supply and demand, or an unusual situation induced by unnaturally low natural gas prices. I don't share that popular wisdom. I think that the low price of electricity, again in my experience in the west, is here to stay for the intermediate future, and I would encourage those who believe all this will pass and things will return to normal, to reflect on the fate of those who bet on a gas bubble that would go away. I think most everybody is still waiting for things to return to normal in the gas industry.

A predictable effect of the low market clearing price of electricity is that the gap between the market price of electricity and the regulated price of electricity is getting wider in a lot of jurisdictions, and that has created predictable desire by consumers to buy at the market price, rather then the regulated price, and to be permitted into the bulk power market. Over the long-term I think these pressures may be slowed down, but in the end consumers will not be denied access to that market. These phenomena obviously create challenges for the demand-side resource industry as it has traditionally been known. The relatively low cost of supply-side options changes evaluation of demand-side opportunities. There is a decreasingly ability of utilities who are in an increasingly competitive environment to pass through the costs of demand-side management programs in their prices, and I think there's a real reluctance in the current regulatory public policy environment for utilities to take on any new fixed costs, not knowing what the future holds for them. I think in particular, investments in demand-side resource programs must appear to most utilities particularly risky, because the value of those investments are so much dependent upon the customers staying with you in the long haul.

Much of the environmental community has reacted with considerable alarm to these events, particularly the events in California. Some normally progressive but distinguished spokesmen for environmental causes have uncharacteristically adopted a "just say no" approach. I don't think that assists in the debate and I think in the end the market forces will prove overwhelming. It seems to me the issue is not whether energy efficiency is desirable, I think everybody agrees on that. I think that the tougher questions in the new environment are, in a competitive electricity market will external stimulus be required to promote energy efficiency, and if it is, how can we apply that external stimulus without screwing up competitive markets, all daunting tasks. I think those who are resistant to the current changes fundamentally distrust markets. I think there's a concern in the environmental community that competition will lead to profligate consumption and a return to the sorts of dumb investment decisions that utilities made in the late '60s and '70s and early '80s. This perception misses what I think is a key point, which is that a lot of the unfortunate events that gave rise to demand-side resource programs and integrated resource planning were a response not so much to unconstrained market forces, but rather a response to the unintended consequences of cost-to-service regulation that I think are now pretty well understood. I think that, without faulting regulators, it's now understood that compensating utilities based on the tonnage of capital they invested was probably not sensible, and I think that it's also understood that a lot of the need for demand-side resource programs came from the fact that cost-to-service regulation has so thoroughly distorted prices. Therefore, I have some considerable level of optimism, to the extent that we reduce the level of regulation, to the extent the market is increasingly driven by market forces. But a lot of what has been characterized as market barriers will take care of themselves.

In fact, I have a substantial level of optimism that while in response to the first question that was posed to the panel but apparently changed, is DSM in or out, I think if you define DSM as the traditional utility subsidized programs, it's probably well on the way out. In answer to the second question, is energy efficiency compatible with the brave new world, I think it is very much so. And I think that those of you with an interest in energy efficiency should be substantially encouraged, and certainly not discouraged, by these turns of events. I think for one thing you can expect that pricing will get better than it's been, which is to say it will reflect long-term marginal cost where it hasn't before. The power market that I see evolving is one where you're going to have a lot more players, you're going to have a lot smarter players, you're going to have smaller reserve margins, you're going to have players taking substantial risks and needing to hedge those risks. I think given the risks that take place in the marketplace, there will be a substantial market for acquiring demand-side resources or demand-side management from existing customers. Consider, for example, that the aluminum industry in the Pacific Northwest is about 4,000 megawatts load there, which I understand can, without any real harm to anything, be turned off instantaneously for 10 or 15 minutes. That's the economic equivalent of a very large pump storage facility, and yet in our existing regulatory environment we've really not been able to capture a resource like that. I think within the arena of controls and a vigorous market, a lot of that kind of thing is going to be able to be captured.

I also think there will be substantial pressure for desegregation of the industry, which is that we will end up with separate distribution and generation companies. I think that will prove to be a healthy trend. Among other market entrants these distribution companies will be hungry to provide new products and services, and I think the economics for a distribution company to encourage energy efficiency will be different than what currently exists for vertically integrated utilities.

And finally, I think the more people like Mr. Dar that come into the industry with fresh perspectives respondent to consumer needs, the better off we're all going to be. I'm not persuaded that the pervasive involvement of the utility community in energy efficiency has necessarily been good for it. Finally, I think the situation will be improved in an environment where the merits of energy efficiency programs will be judged by consumers, not by regulators, not by environmentalists, not by utilities who have billions of dollars sunk into them, and that inherently an industry is stronger that relies on the support of consumer preferences rather than encouragement by regulators.

Ashok Gupta: I have my work cut out to follow the two gentlemen that spoke before me, especially when the issue is framed as clearly as they have framed it. First I want to make sure that everyone in this audience understands the Natural Resources Defense Council (NRDC). NRDC has been an advocate of change in this industry for over 20 years. We're not here to support status quo, and we fully embrace the idea that there needs to be a lot more competition in the industry. The question for us really is, what are the rules of competition going to be, and how is the competition going to be structured? You've got to remember that even the equity markets are fairly heavily regulated in terms of how markets function, how competition works. The electric utility industry is extremely important from a perspective of the environment and our economy, and the rules that will govern it, and certainly the transition towards much greater competition, are things that we all need to talk about. This is not something that you just on one day decide we're going to have competition and there it is after 80 years of a very different set of rules, in terms of cost-of- service regulation. Again, these are rules that we've been working very hard to change, because we don't think the environment has benefited from the old structures. So the question is really, how can environmental concerns be taken into account as the industry is restructured, what are the rules of competition going to be?

Most of you already know that the electric utility industry is responsible for approximately one-third of our NO_X, two-thirds of our SO_2 emissions, and one-third of our CO_2 emissions, so in terms of global warming and acid rain issues, and if you look at the impact of hydro projects on rivers, and if you look at the nuclear issue and radioactive waste, all these important environmental concerns are related to the electric utility industry. This industry is extremely critical and that's why NRDC has made it its business over the last 20 years to be fully engaged to see that it functions in a way that is most beneficial to both the economic goals and environmental goals.

You've got to remember also that energy efficiency, from our perspective, should be pursued when it is cheaper than the alternative, meaning new generation, or burning fuel in existing generation. So we push for applying the economic efficiency argument that if energy efficiency is cheaper it should be invested in, and we do not see energy efficiency investments as a social program. Energy efficiency should be looked at as whether it meets the economic test, and our economic tests include the environmental costs. In terms of meeting our Clean Air Act obligations, we should factor in the cost of Clean Air Act compliance when we look at the supply-side options, and ask "is efficiency cheaper than the cost of generation and the environmental costs that will have to be imposed in order to meet some of the obligations in either the Clean Air Act, or even in President Clinton's Climate Change Action Plan."

The first question was, is DSM in or out? Our argument would be that energy efficiency better be in, and better be part of the program, and should be part of the new rules of competition, in terms of making sure that the rules don't discriminate against energy efficiency. So it's critical from an environmental perspective and an economic competitiveness perspective, that energy efficiency remains a part of the mix and continues to be pursued aggressively, because I think that everybody in this room understands that there is a lot of energy efficiency out there that is cost-effective today and should be undertaken and is not being undertaken.

We have embraced the idea that we should begin with and emphasize much greater competition in the wholesale markets. We believe that a lot of the benefits that people expect and want from competition can be achieved by promoting wholesale competition, including divestiture of generation from utilities. Whatever it takes to make sure that the wholesale markets are as competitive as possible, we are willing to support, because at the wholesale level the electricity market is a commodity market. At the retail market, our view is that the market is a service market, and we do want to see a lot more competition and choice at the retail level, but we want to see that market evolve on the basis of providing least-cost service to the customer. Here again, the bill being more important than the rate at the retail level. And so our emphasis at the retail level is much greater competition and choice to the retail customer, where the choices are really among services that will reduce price and produce lower bills. In the transition, there's the whole issue of transition costs, past investments made by utilities that have to be dealt with. There's the issue of, will there be an aggregator, somebody or a utility, that buys the power and operates the distribution system. Is there some function in terms of integrated resource planning that can take place at that level? We think there is a role for integration and for what we call portfolio management, to make sure that longer term concerns about efficiency investments and new generation can be factored in by a portfolio manager. And we also are certainly supportive of moving away from cost-of-service regulation to more performance-based regulation, based on rewarding utilities on the basis of their performance, not in terms of their pounds of investment in capital.

We are on the side of supporting much more competition. We just want to make sure that environmental values, environmental concerns are fully taken into account during this transition, because this is an extremely critical industry.

Let me conclude by listing the five principles for restructuring that we espouse. One, is to first promote more competition to minimize electricity bills. Two, is to remove barriers to efficiency by encouraging utility portfolio management, that it does not discriminate against efficiency investments. Three, is to break the link between utility profits and sales to discourage wasteful electricity use. The fourth is to tie utility profits to their success in minimizing electricity service costs. And five, is of course to protect the reliability, safety, and universal access characteristics of providing electricity service.

Eric Hirst: Let me start by presenting some data. What is the current situation with regard to U.S. electric utility DSM programs? If you compare utility expenditures on DSM from 1992 to 1993, there's a 9% increase. If you compare the utility estimates as of the spring of 1994 with the expenditures for last year, there's another 4% increase, and if you look at the year 1998, which is the last year for which we have the census of electric utility reports, there is a further increase in DSM expenditures. So, reports that the sky is falling and DSM is dead are definitely premature. But, although there is continuing growth in utility DSM expenditures, that growth is occurring at a slower rate. However, if you look at the energy savings and the load reductions (that is, the benefits of the DSM programs), those are continuing to increase at pretty much the same rate that they had been for the last few years. This suggests that DSM programs are becoming more cost-effective over time.

We are asked what's going to happen to DSM in the future. A great deal depends on what the structure of the electricity industry looks like. If you don't have retail wheeling, that is, if you have wholesale competition only, I think you can have DSM programs that continue, with some changes, more or less as they are today. I agree with George Galloway that electricity prices are dropping. Since DSM is intended to be a cost-effective resource, that means that the barrier that DSM has to overcome in terms of cost-effectiveness is tougher. DSM has to be cheaper and has to save more energy for it to pass the cost-effectiveness hurdle. If you have retail wheeling, then DSM is going to be dramatically different. Using a little bit of jargon, historically utility DSM programs were designed primarily using the total resource cost test, the TRC test, which took a broad societal perspective on the costs and benefits of DSM programs. If you go to retail wheeling, utilities, or more accurately, whatever entities sell electricity and electric energy services to consumers, will have to use the rate impact measure. In other words, they can offer only those programs that do not raise prices to other consumers, or as the first speaker put it, they're going to be able to run programs only if they can sell them to those customers who want to buy. They can't distribute red polyester jackets unless customers are willing to pay for them and pay the full price for them.

So, the key question is, are we going to have wholesale only competition, or are we going to have wholesale plus retail competition? A key there is going to be state regulators. If the state regulators believe that energy efficiency and other societal functions are important, I think that there are a number of steps that the PUCs can take to ensure that these benefits, energy R&D, low-income retrofit programs, and energy efficiency, are maintained. The PUCs could set exit and reentry fees for those customers that choose to leave a system, or as prodigal sons, want to come back to the local system. The commissions could require advance notice before a customer leaves the system or before a customer is allowed to return to the system. The commissions could impose charges on transmission and especially the distribution system, to make sure that even if a customer was buying the electricity from some other entity, not the local utility, that customer would still have to bear the same burden as other customers to help pay for DSM, renewables, R & D, and so on.

Commissions also have a lot of influence over how utilities charge for the use of the wire system, in terms of things like regulating margin, spinning reserve, voltage support, frequency control, and power quality. And finally I think regulators can impose requirements on the non-utilities, brokers and marketers, the other entities that plan to sell electricity and energy efficiency services to retail customers. After all, you need a license in most states if you want to cut hair, so it doesn't seem unreasonable to impose requirements on those entities that want to sell electricity and related services to customers. There will be differences in how DSM evolves according to what the structure of the industry is, whether we have wholesale only, or wholesale plus retail competition. In spite of that, I think DSM is going to change no matter how the industry structure evolves, and what I see in the future is greater emphasis on customer service and customer productivity. And correspondingly, that's going to mean less emphasis on the traditional purposes of DSM programs, that is resource value, the energy savings and the load reductions.

On integrated resource planning, Potomac Electric Power Company, the local utility here in the district, just issued its IRP plan in which it says it can run DSM programs that meet the rate impact measure, so they would survive in a competitive market, and they can still get 70% to 80% of the resource value in terms of energy savings and demand reductions. If that's a generalizable conclusion, that's very encouraging for the public policy concerns. If, on the other hand, the PEPCO result is atypical, then there is cause for concern. I think also we'll see a lot more innovative pricing. George mentioned that reserve margins will decline, and I think the primary reason for that will be much greater use of real-time pricing and other pricing strategies. Those innovative pricing schemes will have a big effect on demand, but have very little effect on overall energy use.

In summary, there are two key wild cards. One is the role that the public utility commissions will play, and at the moment I'm a little discouraged. My impression is that most commissioners are standing on the sidelines and acting as if all this was happening to them, as if they were bystanders, rather than getting out in front and trying to take control of the issues. A notable exception is the Connecticut Department of Public Utility Control, which recently issued an order that, for the time being at least, rejects retail wheeling in Connecticut. The second wild card that I see is global warming. If we as a society decide that the threat of global warming is really a serious one and that we need to act now, we're going to have to do something to protect energy efficiency programs. The President's Climate Change Action Plan, in particular the Climate Challenge part of it, relies on the electric utility industry, or more generally the electricity industry, to provide a lot of the reductions in greenhouse gas emissions, in particular carbon dioxide. Without energy efficiency programs, it's going to be very hard to meet those greenhouse gas reduction goals. On the other hand, if we as a society decide that the scientific conclusions at the moment aren't strong enough to warrant immediate action, that we can wait five or ten years, then it's going to be much tougher to support the kinds of traditional DSM programs that we've had.

Irwin Stelzer: I think we should begin by understanding the essentially fraudulent nature of sessions like this. That's very important. What happens at sessions like this, is that we invite speakers from the public sector, and they always include the best of the breed. We get Phil Sharp from the Congress, Alfred Kahn or Peter Bradford from the regulatory community, Ashok Gupta or one of his colleagues from the environmental community, from which we are asked to conclude that this is a random sample of speakers from those various communities. Not true; it's always the best of the breed. So that when we think about the public policy issues that utilities confront, don't think about the problem of trying to persuade a Phil Sharp or a Fred Kahn or an Ashok Gupta; think of the people who testify before the California commission for positions that are slightly outrageous.

We cannot decide about DSM without understanding the rationale of those programs and examining the relevance of that rationale in today's circumstances. So with your permission, I'd just like to step back a bit to discuss what some have described as a quasi-religious issue. Those of us who have been around this business for more years than we care to admit, ask "Do you remember Reddy Kilowatt? Do you remember nuclear power 'too cheap to meter?' Do you remember General Lincoln?" See, nobody does. Or Lee Metcalf, or Edison Electric Institute meetings in restricted hotels? If you remember those things, you qualify as an industry historian. You would also probably then remember when Guy Nichols, then CEO of NEES, unleashed the first DSM program on an unsuspecting and incredulous industry, in response to the fact that financial markets were telling him they didn't want his company to build any more power plants. This was a shrewd way of saying he got the message and would not encumber his neck with anything like the Millstone being constructed to his south. That explains the earliest basis for DSM: an attempt to avoid expenditures for facilities that would never be recognized in rate bases, and even if included would never be permitted to earn a reasonable return. This early DSM program was part of a very sensible shift in power in the industry from engineers to people who understood financial markets – and I say this with due respect for all the engineers here who just like to build things. That, after all, is what engineers do; when they were deposed in some companies, we had the beginning of sensible DSM management.

As time passed, the arguments for DSM became more sophisticated. Environmentalists helped us to understand that until we had marginal-cost pricing, the economically correct pricing of electricity, we would never get a balance between the demand side and the supply side of things. The price signals that consumers were getting had to be correct. They also led us to understand about externalities, those costs that were not reflected in prices. We were not paying for the resources we were consuming when we threw the switch. This was a kind of a demand-side market failure, compounded on the supply side by the desire of utilities, once a rational desire, to expand rate bases so that they could increase earnings.

Grafted onto these very compelling reasons for demand-side management was the desire of utility executives to woo increasingly green regulators with promises of free lunches for the most numerous customer classes, to allow regulators to engage in that most satisfying of all government activities, cross-subsidization. This was particularly attractive since competition in the telephone industry was making it difficult for them to parcel out gifts to rural consumers, members of the clergy, poor people, whoever they choose, so electricity was the place they had to go to find such gifts. And DSM was certainly useful to purveyors of uneconomic gadgets who wanted to peddle their light bulbs and such.

More important, the desire for DSM programs provided the basis for a new regulatory bargain. The original one was in tatters, sinking under the weight of nuclear plant cost overruns. The new compact goes something like this, and to protect the innocent I've deleted the names of the people from the transcript and I'll just call them executive and regulator.

Executive: Psst. Would you like some DSM?

Regulator: Is that the stuff those pesky environmentalists and my governor want me to try?

Executive: The very same.

Regulator: Is it expensive?

Executive: I don't know, but neither does or will anyone else. Besides it won't cost you a dime.

Regulator: Well, what's in it for you?

Executive: I get to cover my costs plus a modest profit that no one will be able to compute, plus I get great PR and with luck, the Amory Lovins Award for creative construction of statistical and cost data.

Regulator: What's in it for me?

Executive: Quite possibly dinner with Al Gore and a special award if you stay awake during his greeting.

Regulator: Will we get caught?

Executive: Not a chance. Some fringe group of industrial customers will try to expose us, but they're interested in profit making and therefore have no credibility in regulatory or political circles.

Regulator: OK, let's give it a try.

And we got DSM. Now it was not a bad thing, at least at first. Uneconomic pricing, imperfect information, industry lethargy, all were imperfections that had to be overcome, and some dose of offsetting artificial manipulation of the market might have been necessary to do that. Or at least a case for such off-sets might be made.

The question that remains is, are the original reasons for this exercise still intact? I don't think so. Recall, when we started DSM there was reason to believe that average costs were below marginal costs so that people were over-consuming. Now marginal costs are below average costs, and so there's no reason to believe that people being charged average costs are over-consuming. Recall too, that there were uninternalized social costs at the time. We now have good reason to believe that most (or perhaps more than all) of the social costs have been internalized, and that if there are any uninternalized social costs left, they can be internalized in a more efficient way than with artificial DSM programs. Recall too, that when DSM was introduced there was widespread belief that consumers were too ignorant to manage their own electricity use, that there were information gaps in the market, and so on. I think that the response of consumers to the run-up in gasoline prices during the various supply interruptions and their response to the lower prices now, shows that consumers are pretty well tuned in and able to define their own interests. Also at the time, there was a question of whether or not utilities had a huge incentive to build things, and therefore to discourage DSM programs. I think unless we let them off the hook on the question of stranded assets, and by that I mean plants that they built when they shouldn't have, not contracts that they signed with a regulatory gun to their head, there's no reason to believe that they will continue to have an incentive to build things that they shouldn't. So we're left with very, very little need for the demand-side management programs that are threatened by competition.

Which brings me to my conclusion and to my answer to the questions put before us. Are there market barriers? No, not as traditionally defined. I think that one market barrier that remains is the vertically integrated structure of the electric utility industry. George assures me that his clients are not wedded to that and therefore, will be delighted to become energy service companies and find someone to buy their generating plants. That would add to the portfolio that the person had which would include the Brooklyn Bridge. How do you deal with these imperfections? You don't have to, they're not there. Rates versus bills: rates matter, they determine consumption on the margin. What structures will evolve? Vertically disintegrated ones. Can DSM fit into a competitive model? Yes, it is entirely consistent with a competitive model if the programs are efficient, as Mr. Dar argued. In fact, all that competition does is doom inefficient DSM programs and other atrocities such as California's forced purchase programs. It is astonishing to me, by the way, to see how willing DSM advocates are to admit that their programs are threatened by competition and therefore, to applaud for example, the Connecticut Commission's decision to stick with Soviet-style central planning. I had to say that because Eric said "Please don't say Soviet-style central planning." Third, I think competition will democratize all conservation and environmental programs, by making the trilateral conspiracy between utilities, environmentalists, and regulators very difficult to maintain in the future and move these issues into the political arena, where they belong. These are taxes, they should be decided in the political process by legislators. We know how fond legislators are of levying new taxes, now seen as a predicate to retirement. And I think competition liberates sellers of DSM programs from the utility ownership and domination that they now have. I would like to see all those programs spun off into viable companies competing in the marketplace with anybody else who wants to sell such services. Whither as for IRP, yes it will wither.

Questions and Answers:

Marritz: I reserve the right to ask the first question. And let me aim it at both Vinod and Irwin, and the parting shot about wither IRP prompts the question. Can you really write off IRP? If in the model Vinod talks about retail customers cannot go to the electricity supermarket unless you have retail competition, and you really have customers having free access to go and pick and choose the products they want, doesn't somebody need to assure that the local utility, still a monopoly, needs some accountability, has some accountability?

Dar: I view IRP as sort of sensitivity training for the prison guards, and frankly, just because you make the prison guard more sensitive doesn't mean it's still not a concentration camp. So my view is that whether we have imperfect competition or perfect competition, the real debate is not that we have imperfect competition and therefore we need sensitive prison guards, the real debate is, can we continue to afford the cost of absurd monopolies and therefore satisfy ourselves that because we've shown the prison guard videos on equal opportunity and sexual harassment, that somehow we have addressed the fundamental questions. IRP was a bad idea to start with. It is fundamentally incapable of working. Hayek basically settled the issue 40 years ago, I advise people to go back and read that. Anyone who's read Hayek will understand you can't do IRP, it's impossible.

Hirst: To say the least, I disagree. I think IRP has accomplished a great deal in the ten years since it has been in existence. I fail to see any analogy between planning for the electricity industry and prison guards. Let me turn it around. Every time you flip the switch, you're setting in motion a chain of events that go all the way back to a power plant, through a very complicated grid and control center. And when you flip the switch, it affects everyone else's use of electricity: power quality and reliability. There is no way to run an electric grid without having some central planning. I invite for your consideration Chicago O'Hare Airport where we have eliminated Soviet-style central planning, and you Irwin, can fly your plane in, you can land any time you want and you can take off whenever you want. We're going to have a lot of crashes, and if we don't have some kind of centralized planning and operation, the system is simply not going to work.

When you have a more competitive environment, the nature of that planning and who does it is definitely going to change. If you have a retail competitive environment, all the traditional IRP planning functions will remain, but they will be carried out by different entities. The generating companies will do generation planning, the transmission and distribution companies will do transmission and distribution planning, and the customers who choose to will do their own resource portfolio management. But somewhere there's got to be some entity that makes sure that the system is in balance second to second.

Stelzer: First of all, consider the notion that somebody has to do planning. If you have retail competition, someone has to plan generation, in a certain sense. What he has to do is plan how best to meet what the customer says he wants, which is a different thing from sitting down and planning what he wants to give the customer, and letting the customer figure out how to use it. Second, it is indeed the case that anyone using an electric system, anyone using an airport, affects third parties. Every transaction, every contract in society affects third parties; that's why we have arrangements where third parties get reimbursed if their property rights are adversely affected by a deal between two parties. You can't have two parties make a deal that squeezes somebody off a transmission system without reimbursing the person who's squeezed off the transmission system. So sure, there will be more complicated transactions involved.

But think about IRP. You have utility executives who have never met a market, and probably never met a customer, because in most urban areas they probably try to live outside their service territory. They've never met a customer, they've never met a market, they've never seen a market-based price. They sit down with risk-averse regulators who themselves have opted for a career that avoids market competition. All those risk-averse people sit down and decide how to plan an electric system. I don't think the electric system here will collapse if we continue with such central planning. It will simply be sub-optimal. It has been sub-optimal for years. We now have a chance to make it what consumers want it to be, and we're clinging to a whole bunch of stuff as a mask for social objectives that have nothing to do with the consumer of electricity.

Dar: I would like to very briefly and respectfully respond to Eric. The fear that he raises, that undisciplined and disorganized behavior will lead to accidents, is quite correct. But markets, especially markets where many transactions are done very quickly, are masterfully self-organizing mechanisms. You could buy a stock today in Singapore and affect global capital markets everywhere. Financial markets have created clearing mechanisms, hubs, transfer mechanisms, standardized transactions. Natural gas markets and telecommunications markets have done the same thing, because it is profitable to organize and clear transactions. In the electric industry similar market clearing, transaction clearing, if you like, traffic control and dispatching organizations, will emerge. Just because we have traffic control at O'Hare doesn't mean it can't be done by a private company. Secondly, with due respect to Professor Stelzer, what is sub-optimal to the professor, is a going out of business sale for a real consumer in California.

Q: What happens to the nuclear power plants that don't compete, and who operates them during the transition, deals with the safety issues, and closes them?

Stelzer: Well, the ones that have variable costs above the average total costs of new plants should be closed. That seems to be clear. I think the companies that erected these things should have the responsibility of removing their trash. The notion that this will lead to huge write-downs, by the way, is absurd. The market has already written these assets down. The only question is, can you get your accountant to move his pen to reflect what the market already says. These are unfortunate investments. The utilities made them, many of them at a time when they knew the social compact to be frayed. Nobody forced them to do it. They fought like hell for the right to build these plants. That's different from these contracts to buy more and more power at 16 cents, so beloved of the California Commission, which forced the utilities to take that power.

Marritz: Also under the NRC license, I think the licensee is obligated to do all the things he's obligated to do. Decommission, dispose of fuel, and so on.

Gupta: From NRDC's perspective a much more competitive market has benefits for the environment. Clearly, in terms of the nuclear issues, a lot of dirty coal plants, a lot of dirty plants out there should be retired, and a lot more clean gas should be coming in that's much more efficient, and much cheaper. So from an environmental perspective, competition does not have a big downside. The question again is, what happens at the service side and the customer side of the equation? We believe through wholesale competition we can achieve a lot of the environmental and economic benefits we want, without losing the huge potential that still exists on the customer's side of the meter. Here the question is the energy service companies. Through competition you can get energy service companies to provide a service to the customer, and we have always supported that. The DSM efficiency industry is going to evolve and we're going to get smarter and be more cost-effective at achieving the efficiencies. It's the direct access issue that's still the most troubling from the point of view of achieving some longer-term objectives. We want to make sure that the portfolio manager takes into account the benefits that can be achieved through longer-term investments in efficiency and renewables.

Stelzer: Why would that be? Why would that market failure exist?

Gupta: For example, wind is an intermittent resource. It makes best sense combined with other resources and being part of a portfolio, rather than having to compete purely on its own.

Stelzer: But why wouldn't some aggregator put together such a portfolio?

Gupta: Right. Our question is, who is that aggregator going to be? During the transition there isn't room for a distribution company. To the extent that there is still a monopoly function in terms of the distribution company, we see no problem with having the distribution company being the aggregator. So our positions are not really that far apart, in terms of how to achieve these environmental and economic efficiencies. It's not a matter of turning it into a pure ideological battleground. Because it's not about ideology.

Stelzer: Are you more comfortable working with a monopolist to get a stated objective, then letting a market develop? That's disturbing.

Marritz: Let me come back to your statement, Irwin. You said you don't see why we need an aggregator.

Stelzer: I say we do. I think some guy will appear who will put together this portfolio. Stocks is a great analogy. What happens with securities? People put together portfolios, they call them mutual funds and sell them to people, and they've got a variety of risk-reducing or risk-enhancing devices. I think you're real problem is that Ashok doesn't like the discount rates consumers are going to apply if they have choice, because consumers always have higher discount rates than intellectuals. That is, intellectuals always figure out, "Well, it'll pay in seven years and consumers didn't buy it, so they must not really know." No. Consumers know they're going to move every three years and they don't want to put in the insulation that's going to pay in seven, because they're out of there. They know their own discount rates, and most intellectuals and planners don't like consumer's discount rates, they're too high. And I think that's too bad.

Q: Listening to the panel there are, of course, very different points of view, but I do see something in common, which is a much more selective, careful look at DSM than we saw several years ago. How might this affect our frequent advice to other countries, in our aid programs abroad? I think we've pushed DSM quite strongly as a method of introducing energy efficiency, and I would like to ask whether they think this is the moment perhaps for introducing a little more caution in these programs.

Dar: I guess giving advice to other countries is a bad habit we all get into, but it seems to me that each country is so different. And within each country, each region is so different. There may well be countries where you work, where it is simply not possible to move surplus electricity from one part of the country to another, or be connected to a large grid. In which case, the marginal cost of producing and transmitting power is going to be extremely high. Under that circumstance, efficiency may be the best thing going. In other instances, if DSM has been pushed because it provides opportunities for the affiliates of U.S. electric utilities who can't make money and have to go start a business elsewhere, then I think that's the wrong motivation. But by and large, if we are using large companies as experts, I don't think they would be pushing solutions that are biased in one direction or another. As electric utilities restructure internationally, the debate we are having here is no doubt going to be replicated over and over again. So I think what we should be advising people on, is our experience in moving form a monopoly to competition, as opposed to advising people on outcomes.

Hirst: We've learned that, in general, DSM programs save less energy than we initially thought they would, and they cost more than we initially thought they would. But overall, relative to the total resource cost test, which Irwin and George would reject, the programs are still cost-effective. DSM programs can contribute, they are cost-effective, at least by those traditional measures, but not as much as we initially thought.

Gupta: My answer has two parts. One is to clearly focus on making the smartest economic decision. If efficiency is cheaper than building power plants, make those type of investments. From a consumer perspective or from an investor's perspective, use the economic test and make the decision so it still might make sense. You shouldn't assume anything about whether DSM makes sense anywhere, unless you do the economic analysis and see which comes out the winner. In terms of Eric's point that some DSM programs work better then others, there is a lot more experience now. A recent study came out of California's '90 to '92 programs, and the actual measured savings from DSM were estimated to be $2.5 billion. There are different programs, some have overachieved, some underachieved, but the total savings from DSM programs invested in by the California utilities was actually not far from what was originally estimated.

Marritz: I might add, with respect to giving advice to other countries, I didn't hear any of the panelists say that DSM or energy efficiency, was a bad idea or would not and should not be practiced in the future. Indeed, Vinod, if I heard him correctly, said that he thought it would be delivered more effectively by the market than it ever was by the utilities.

Stelzer: That problem is really that there are huge environmental resources in which there are no property rights, and we're talking here about the distribution of those between us and them. When we have programs that we think ought to be attractive to them, it is usually designed to enhance our access to those unclaimed property rights and vice versa. These are huge questions about international distribution of income, and the programs that I've been involved in where they talk about energy efficiency, they are really all subsets of questions about redistributing income north to south. I think certainly, the idea of models being available and spreading information is fine, but you can't attach the same numerical values to the programs in other countries that need economic development much more than we do, and on the margin care less about pollution then we do. You can show them the programs, but the values assigned to each of the things have to be indigenous to the countries involved, and you always have to be sensitive to the fact that you're engaged in the international redistribution of unclaimed property.

Q: Given the fact that the world has a large number of very large environmental problems, and given the fact that we don't have a whole lot of ideas other than efficiency improvements and renewables to do something quantitatively significant about these, I think it's very appropriate to ask questions about how we arrange these systems so that we reach overall global objectives. Looking at the advice that has been offered from the panel, it's largely based on improving efficiency and competitiveness, and I think that is a very good approach. However, if we look at the situations in other energy markets where competition does exist, we're not seeing that the efficiency improvements are being captured, or that new technologies are being brought to these markets. So that raises the question, is there perhaps anything more that needs to be considered, and if so, what?

Stelzer: Could you give us an example of the market you're referring to where we have competition but not efficiency?

Q: Well, it seems to me that in fuels we have a lot more competition then we have in electricity. The petroleum markets for example.

Dar: I disagree with the premise and the conclusion. In natural gas and oil, where competition has been very vigorous, we have seen dramatic increases in efficiency, for example, of the drilling dollar. We have seen dramatic increases in the efficiency of threading technology and seismic technology. We have seen dramatic increases in the productivity of coal bed methane extraction, tight sands extraction, and in fact today, the drilling dollar is probably twice as productive as it was when we had controls on the industry. And the reason, of course, is that in a competitive market it pays to deploy efficient technologies, because that is how you gain an edge over someone else, especially when the long-term price of your commodity is falling.

The second point I would make is that in contrast to just four or five years ago, today it is possible in the U.S. for an independent competitor to deploy a state of the art combustion turbine, using current gas prices, heat rate 7,500, and produce power cleaner, more efficiently, and deliver it for less than 4 cents into the grid, compared to combustion turbines in New York, where people are looking for 6 cents as floor pricing. So I think we have seen some fairly dramatic increases in the production of energy, because those are the markets that have been the freest. In the gas transportation industry, the transportation network is certainly now a lot more productive than it was before competition came to gas, and my guess is that when competition comes to consumption technologies, there'll be very substantial increases in the rate of capital performance.

The related question that was, if all we can come up with is increasing the productivity of capital, given the hypothesis that we face these vast environmental problems, then is that good enough? And the answer is absolutely not. But why do we have large environmental problems? We have large environmental problems where markets are not allowed to work, and we have small environmental problems where markets are allowed to work. Therefore, those who truly believe in cleaner air, cleaner water, cleaner everything else, ought to really be pushing for property rights. Environmental waste is economic inefficiency, and you will have environmental waste as long as there are no price signals. So if we could get free competitive markets in air, in water, in noise, in garbage, people would learn to use those resources efficiently and many of our environmental problems would decline. It is not an accident that the centrally planned sites are the very dirtiest.

Q: I'm not so sure that all these efficiency opportunities are being captured. The example you gave is basically on the supply side, that gas turbines have become much more efficient. However, if you look at where the opportunities exist, you have to look at the demand side, and then the situation is totally different because of the structure of the old setup. You may hypothesize that there are no global environmental problems. I think the overwhelming scientific evidence is the other way. And there's also the political aspect of it that perhaps is relevant, that more than 150 countries have signed a convention on climate change. That, if nothing else, has political implications, and for a country which emits 25% of the greenhouse gasses, it may also warrant some discussion.

Gupta: I think the same technology improvements and efficiency improvements we're seeing on the supply side are there on the demand side. The question then goes back to the issue of barriers, and what is it that prevents improvements from really coming into the marketplace, and how can you best overcome the barriers? There are still some barriers on the demand side. We certainly are working on refrigerators and other appliances, we know that the manufacturers can produce appliances that are much more efficient, but the manufacturers believe that consumers will not pay a slightly higher up-front cost to purchase those appliances. Can they compete in the marketplace, and what is it that keeps a more efficient appliance from being purchased? What about the financing issues and delivery mechanisms, because this is something you have to deliver in each person's home, versus going out and building a new power plant. How much harder is it to deliver a more efficient appliance and market it? It is a harder thing to do, but we believe the efficiency opportunities are just as great as on the supply side, and what we're trying to figure out is a way to capture that.

Stelzer: Doesn't it strike you as just a tad strange, that the NRDC sees more market opportunities for types of refrigerators than Jack Welsh does at GE? And as to the other point about the 25% of whatever the U.S. produces, it also produces 25% of the world's GNP. So those are proportionate numbers, that's about what you'd expect.

Gupta: Well the answer to that is maybe not GE, but Whirlpool clearly appreciates it, and Whirlpool has struck out in terms of making the investments, not only in these more efficient refrigerators...

Stelzer: People didn't want it.

Gupta: That's not true. Consumers need choice is what I keep hearing, and the question is, are consumers really getting the full range of choices on energy efficiency?

Q: The question I want to raise is one that's been pervasive all morning, and that's the change from the ethic of conservation to efficiency. The question, I think, may be naive, but the more I listen to the discussion, the less naive maybe it is. And that's the difference in efficiency as thought of by engineers and economists.

We're talking about either economic values, in terms of dollars, or we're talking about resources, in terms of BTUs or other energy units. And there are many different ways of thinking about efficiency. I want to propose that we come up with some kind of common definition, if we're going to use the term efficiency. Secondly, Mr. Gupta mentioned that we ought to consider energy efficiency if it's less expensive then new resources, and I'd like to hear comments on that from the panel.

Hirst: I agree that there are some semantic problems, that we talk about conservation, about energy efficiency, and about economic efficiency. Here on the panel, we've all been talking about economic efficiency. What divides us is our perception of the existence or nonexistence of market failures that prevent customers from adopting energy-efficient technologies. Some of us here think that markets work well, and that there aren't any serious market failures, let competition prevail and what will ensue in terms of energy efficiency is economically optimal. Some of us think otherwise, and believe that there remain market barriers. Ashok mentioned one just a minute ago, that customers often don't have the choices they ought to have in terms of a range of efficiencies, and so they are not even allowed to make choices that would affect their energy use and efficiency. So we see a continuing need for some government intervention to help overcome those market failures.

Galloway: My take on this is, we need to ask ourselves why is it the case that in every other good and service produced in the American economy, we conclude that the right amount of consumption is that which consumers elect to buy in a freely competitive market, but for some reason in regard to electricity, we should mistrust that consumption decision and there should be some external engineering analysis made of the "right amount of electricity to be consumed." That may be the right thing to do, but I think that where that departs substantially from what we do otherwise in the economy, it carries a very heavy burden of proof.

Gupta: A quick answer to that is that there are huge environmental consequences at stake. I really do believe with the electric utility industry, unlike the telephone industry where residential consumers are probably using the phone 20 minutes a day, if people doubled their telephone usage, there could be greater economic benefits, but if all residential consumers doubled their electricity consumption as a result....

Stelzer: Even if externalities are internalized?

Gupta: No, not if externalities are internalized, but they're not. Maybe everybody agrees that externalities should be internalized, but then my response is that we would rather see that happen first, before we give up on the role that utilities can play in helping remove some of the market barriers to energy efficiency. The economists can debate that point.

Q: Let me preface my question by saying, without apology, that I too am an engineer and I do believe in the free market, so don't get me wrong. But I'm looking for answers. If the free market drives down the price of electricity, is there a chance that the externalities may remain external, never become internalized? I equate carrying out the trash with dealing with these externalities, and will there be money left in the budget, if the profit margin becomes so low, to carry out the trash and prevent future accumulation of trash?

Stelzer: The answer is yes, if you get the rules right.

Dar: The issue is not the level of the price, but how things are priced; it's what you put into it. Price contains all the information you would possibly want to know about what you want to do with something, as long as it's priced properly. So I think the debate shouldn't be about, if competition drives down price, then will we be able to sell expensive engineered solutions to people? If I were an engineer and good at it, I would spend my time responding to price signals, producing increasingly innovative and efficient technologies that people want to buy because it increases the productivity of capital. And as long as you are in the business of increasing the productivity of capital and delivering value to consumers, you will have an expanding market.

Gupta: As an economist I think I could probably get away with saying it, the old joke about the economist knowing the price of everything, but the value of nothing. I think there is an issue of price versus value here.

Stelzer: As long as you separate them for analytical purposes, I think that we can come close to agreement. And the previous gentleman's question: if you have some time someday, go to the library and look up the sumptuary laws that prevailed in Italy in the 14th or 15th century, where they had notions of how many chickens you ought to eat at a dinner, and how much braid you should put on your costumes and so on. That, carried to its extreme, is the same as the notion of how much electricity one ought to use.

THE DIFFICULTIES OF DEREGULATION: CONSUMERS, STRANDED COSTS, AND THE PUBLIC INTEREST

Llewellyn King, Founder and Publisher, King Publishing Group.

Conservation and efficiency have become, of course, apple pie and motherhood. They're all terribly good goals, and they have been ever since the energy crisis of 1973-74, when it became blatantly clear that we had to reorganize our affairs in energy. We have come to have a regard for the environment in every action taken in industrialized societies, and in less industrialized ones. It is an enormous achievement that a concern for the environment, a concern for natural resources, has been institutionalized in human conduct. This does not mean that the game is won. The production of energy has an environmental impact no matter how it is done. It will be, for the rest of mankind's history, a continuing struggle to look out for the environment, therefore to be more efficient, and therefore to be able to extend, to the great many people on earth who have never know the quality of life we have know in the West and in Japan.

The United States is in one area especially remiss in conservation and efficiency. We, as public policy, believe it is essential to keep the cost of gasoline low. This is a very contentious subject. We send a signal to people to use gasoline, and to consume an enormous amount of gasoline to accomplish a job that less could do. But we have never had the political will to go after the enormous use of petroleum. There are too many stakeholders for a gasoline tax it to be politically popular. I believe to this day that if we increase substantially the taxation on oil products, outside of their essential uses, which include diesel and farming, we would cut our import balance, we would help the environment, and we wouldn't hurt anyone.

It has been easier to work on other energy sources, especially the electric utility business, sitting there large, conspicuous, measurable, with no real constituency save itself. Ratepayers of electric utilities do not feel that they are stakeholders: they do not feel responsible for, or ultimately affected by, what happens to that utility. We tend to think that the price of electricity is an act of God, and if the politicians add some taxes on, so be it. And yet, as individual consumers we generally have less choice and less elasticity with electricity than we do with petroleum products. This is an odd dichotomy. The greater national good would be done by going after petroleum.

The crux of the current challenge is, how do you deregulate the electricity business? Messily, obviously. Deregulation is very hard. It is not smooth, it is not organized, and deregulating one industry is not analogous to the earlier deregulation of another one. There are bits of analogies for the electric industry, with telephones, natural gas, the airlines, but no one analogy that applies to the whole thing. There are limits to how much commonality there is between the experience of these industries and electricity. The big difference is the real-time, on-demand nature of the electric power system.

One of the least compelling arguments for deregulating electricity comes from my friends in the environmental movement. It seems to be a nostalgia for a state of balance, for a state of loyal opposition, for knowing where your designated villain is, and also for a certain amount of social management that arose in a relationship between the utilities, their regulators, and environmental movements, and spawned demand-side management. On the whole, utilities went along with DSM because it sounded good, and because they could get political and environmental approbation for doing it, although there were less convinced of its efficacy. The public had very little say in it. The utility commissions went along with it for many of the same reasons, and in the hope, but without the conviction, that very difficult new generating decisions wouldn't have to be made.

In the new world, it doesn't seem to me that there really is a place for DSM as we have seen it, because the new brutal arrival on the electric scene who is going to change it all, is price. We in the U.S. actually have very little idea of the true price of the electricity we produce. The New York Mercantile Exchange is working on a contract, and suddenly we will have a published price for electricity, which will be the clearing price of the futures market. This published price will highlight the variations in price that consumers pay based on where they happen to live, and this will be troubling since they tend to take the price of electricity as an act of God. When you move you consider the cost of housing, but you don't consider the cost of electric power. In this new world, the public is going to want a transparent price.

This is not an easy deregulation. There are other things that have not been resolved, and there is no guidance. It is happening almost without the public's knowledge, in the dark, not by conspiracy, but because it is so complicated. Yet it is going to profoundly affect the infrastructure of the nation. There has been more debate about more modest issues in other countries than there has been in the U.S. about this great issue.

There are two components that I think need resolution. One is that for the residential customer, traditionally subsidized, traditionally the greatest single beneficiary of the utility, always the child and the dependent of the utility, it is yet unclear how that price and service will be guaranteed, how that subsidy will be removed without huge pain. In a competitive market the large customer will buy the cheapest electricity, and there will no longer be a preference for the residential consumer. That is the great unresolved social question, and I have not yet heard a good answer. The second thing is the enormous question of stranded investment. There are two sets of opinions here. One comes from pure economists, who tend to say, "tough luck, you built the wrong machine, or you signed the wrong contract." A steadier approach is from the people who built the machines or signed the contracts, who say "we didn't do it, public policy made us do it. We build nuclear power plants because public policy encouraged it. There was an implied threat that if we didn't utilize this technology, we might lose our franchises." This argument is a little disingenuous, but there is some truth to it: by and large we all believed we needed a lot of nuclear, to meet a demand growth that didn't occur. Efficiency came in instead. Now, with greater efficiency about to be introduced by the market, in a ruthless way that only the market can bring about, with lower priced electricity available, these plants are not very satisfactory investments.

The other, and larger chunk of stranded investment, is qualifying contracts that were entered into to purchase power, sometimes from exotic producers over very long periods like 30 years, as a matter of public policy. They were encouraged by the utility commissions, encouraged politically by the Department of Energy, who all exhorted "go for it, alternative energy is the future." The utilities were under political and regulatory duress to enter into these contracts, and they had nothing to lose. The costs were passed through. Their efficiencies were not in the cost at which they bought this electricity. These involve a very large sum of money: it is calculated at something like $200-500 billion. If that's the real cost of stranded investment, somebody is going to get very hurt. This is a national problem, not dissimilar from the savings and loan situation. Once there becomes big political sentiment to do something about stranded investment, there will be a big witch hunt for the villains, and the villains will be the people who did what was perceived at the time to be prudent things, well within the expectations and directives of public policy. Most of the proposed solutions involve a tax on transmission, but we don't know if this would hold up.

My expectation is that there will be political preparedness to assist utilities with stranded investment if that investment is stranded in qualifying facilities: expensive generators, like geothermal, early gas, or wood, things once called appropriate technology. If these were bought under political pressure, the political system will save them. If, however, that stranded investment is purely nuclear power, it is unlikely that there will be a political resolution, because the people who hate nuclear power hate it illogically and pathologically. They are politically adept people, and they not only want to see nuclear power banished from the face of the earth, they want to see those who brought it to the face of the earth punished. It is likely that a large number of our utilities will enter into bankruptcy, come out reduced in size, and enter into court settlements about these contracts and about nuclear power plants. Already Wall Street is viciously discounting utility stocks, and this is very serious because these are infrastructural companies that are getting into very deep trouble.

We're in for five messy years, and a period of great difficulty, at the end of which, because we will shop for electricity and the price will be more critical, the efficiency with which it is used, dispatched, and generated, will go up. The first big gain will be efficiency in generating. What was once considered baseload may be mothballed. New entrants, especially gas turbines, will provide the immediate electricity of the future. It is unlikely, in the foreseeable future, that anyone will build a large, central station power plant of any kind. This is not for an economic reason, but for a public reason: that is deep-seated public concern over electromagnetic fields. Chances are, if you built a plant of more than several hundred megawatts, you might never get permission to put a line into that plant to take the electricity out. A whole set of things is forcing us to diversified generation and diversified ownership, although I tend to think that in twenty years the economists will be back telling us they should all amalgamate, it will be more efficient.

Despite all these difficulties, and the highly political nature of the issues, we all stand to benefit in our high purposes of efficiency and conservation and environmental protection, and in the lesser purpose of our own endeavors to make a living in this world. There are good reasons for people in the utility industry to be anxious, but it is a misplaced anxiety focused on fear about what will happen to specific companies in a free market. They can't keep 100% of their market share. They would do better to think about which bits of the market they want, which bits can they serve well. This is not downsizing, where hundreds of thousands of jobs are being lost. This is a diversification in an industry where there will be more and better jobs. In the long-term the goals of environmental protection and consumer service will be well-served by deregulation, but it is not going to be reached easily, and there will be winners and losers.

PROMOTING ENERGY EFFICIENCY: COMMAND AND CONTROL VS. MARKET PULL

William O'Keefe, Executive Vice President, American Petroleum Institute.
Hans-Eike Von Scholz, Head of the Energy Technology Strategy Unit of the Directorate General of Energy of the European Union.
James B. Sullivan, President, World Energy Efficiency Association.

William O'Keefe: I want to take a few minutes to set the stage for this discussion. Everyone agrees that using energy efficiently is good for the consumer, good for industry, and good for the entire economy. It is always easy to make the assumption that more of a good thing is better than less. But economists know that resources are scarce, less than wants, and that there is no such thing as a free lunch. In discussing energy efficiency we need to know where we've been, where we are, where we want to go, and whether we want to get there by reliance on markets or government mandates.

The U.S. economy is highly industrialized, dispersed, and energy intensive because an abundance of low-cost energy and an economic system based on private entrepreneurs allowed energy-intensive industries, like aluminum, papers, chemicals, and refining, to flourish here. American transportation needs are great due to the long distances required to transport goods and satisfy the demand for personal mobility. But when you adjust for these differences, it comes as a surprise to many people that we're about as energy efficient as Germany and Japan. The U.S. is more energy-intensive than some industrial nations, but that does not mean that we're less efficient. Still, we have made great progress in the last quarter-century, with energy prices being a big motivator. The energy needed to produce a dollar of gross national product has dropped by about 32%, and passenger cars that averaged thirteen miles per gallon in the mid-1970s now average over 21 miles per gallon, a 60% increase. I'll leave it to the panel to debate whether that improvement is primarily the result of price signals or the CAFE standards. Since 1979 our economy has grown by about one-fourth, and energy consumption has remained relatively constant. About three-fourths of the efficiency gains have been due to technology.

Improvements will continue as more energy-efficient investment gradually replaces the nation's existing capital stock. New office buildings and cars will replace old ones, more efficient furnaces will replace older ones, new technology will lead to changes like telecommuting and video conferencing. But it will take years of investment to completely transform the economy. So how far do we need to push energy efficiency beyond the normal rate of progress, and why? How much time is reasonable to achieve current goals for a less energy-intensive society?

Take the case of oil. Government action has been urged to correct perceived externalities. We were running out, supplies were insecure, we were generating too much air pollution, we're causing global warming. Well, world reserves of oil have increased 50% in the last decade, and we now have a 50-year inventory at today's consumption rates. Cars and fuels will soon be 97% cleaner than they were a few decades ago. Air quality continues to improve. Imports are rising, but are less of a threat than they were in the '70s. And the case for climate change is anything but settled. So does it make sense for the government to move the U.S. economy away from oil, or to force the development of new technology, by second-guessing the market? Does anyone remember the Synthetic Fuels Corporation?

I hope in this discussion some attention will be focused on the difference between what government can do, and what government should do. I think government should have two basic goals: to harmonize conflicting interests into social priorities, and to see that externalities that represent true market failures are internalized. We'll be discussing two approaches today. Market pull relies on the government to set goals, address externalities where they exist, assist with the dissemination of information, and set ground rules by which the market operates. The alternative, command and control, places government in a more directive role, determining not just socially desirable outcomes but also how they are achieved. In discussing these issues we should keep in the mind the dramatic changes taking place in the U.S. economy today, as a result of information technology. Our economy is vastly more complex, yet vastly more productive, than it was a decade ago. The information-age economy has been described as similar to an evolving ecosystem where continuous adaptation and competition create productive systems. In this new climate, the issue of what government can do or should do becomes even more relevant.

Hans-Eike Von Scholz: I am happy to be able to talk to you about what we are doing in the European Union (EU) now, in a more integrated manner, on this crucial issue of energy efficiency. The topic of this panel, command and control versus market pull, is precisely the issue we have been dealing with for 15 or 20 years in the energy sector. We have very different economies and systems in the European Union. I remember eight years ago, when Portugal joined the community, electricity prices there were very low. We had a good plan worked out since 1973 for implementing energy efficiency measures, but it was almost impossible to launch them in Portugal when the price of energy was so low. This is similar to the situation at the moment in Central and Eastern Europe. In Greece, electricity is heavily subsidized in some areas. There are large differences within this large continent, even within a rather integrated system. Our energy sector is not nearly as integrated other things, such as our agricultural sector, but we are working hard, and I will talk about two or three areas where we are attempting to act on energy efficiency.

To make my own position clear, I would like to correct the title of this panel, and say, command and control in service to, in order to enable, market pull. This more accurately reflects my position, and I think the position of the European Union. In fact, we believe that the market pull is essential, but in order to work, we need a market. We need a market that can pull, that is functioning, and here you may know that the situation in Europe is in great part worse than in the United States. We have in many areas no market in many sectors, at least no market as we would like to see it. So the very first thing we need from the public sector is to try to increase transparency. The second credo is to provide the adequate legal basis for whatever we need to do: setting standards, regulations, minimum performance standards for heating systems, combustion processes, burners and boilers in industry, etc. All these things are of vital importance and need some public action. The third aspect is incentives. We try to combine these actions as much as possible. But we have different departments, as you have, for environment, energy, internal market, industry, and these departments do not always work nicely together, so you can imagine that it is not very simple to get these policies through.

Let me mention a couple of things about each of these issues. I'll start with the legal framework, for this is an area where we are very active right now in the EU, in order to get all the different legal frameworks of twelve, soon fourteen countries, together. We have been pushing through deregulation, or unbundling, against the quite vigorous resistance of big electricity companies, to free up access to the grids to transport electricity, or gas, within the EU, so that big industrial consumers can buy them more easily. We are not at the end of implementing the whole thing, but we are rather advanced, although there is still a long way to go. This is where command and control can be useful concerning the larger establishment and functioning of a market deserving of the name.

Regarding incentives, there has been discussion of indirect methods, which means taxation and pricing, and here the proposal for the energy-CO_2 tax has been on the table for about a year and a half. One of the main reasons we cannot get any further in implementing it is that our industry is crying that if the EU does it but the U.S. does not, than they would be at a competitive disadvantage. We are looking with much hope to your new government. Unfortunately, for the moment the CO_2 tax is out of the discussion.

We have rather big programs of direct incentives, like you do in the U.S., in particular for innovative technologies. I think you have bigger incentives from your environmental agency than we do from ours. In the past four years we have spent $1 billion for research, development, and demonstration of energy efficiency, renewable resources, and clean use of fossil fuels. The European Council has just decided to authorize another $1 billion program for the years 1995-1998. This is significant seed money to push technologies where the market would otherwise not take them up. These are technologies where either the economics are wrong, because in Europe too the internalization of externalities has not progressed very far, or we need these types of incentives because the transparencies and lack of functioning markets are so great.

Regarding transparency and better information, four years ago we set up a big network of agencies, called the Organization for the Promotion of Energy Technologies (OPET). It now includes 50 agencies, and not only do they get money from the EU, but they also work for their home governments, and the joint efforts of what they are doing for their own states and for the union makes for an optimal system of exchanging information. We resisted setting up a European Union energy agency, deciding instead to work through these national agencies.

To conclude, I think we need a very good cocktail, adapted to each region or state where we are working, a cocktail of using regulation on the one hand, and using incentives on the other hand, in order to make the market work as well as possible, and to generate this market pull.

James Sullivan: I would like to reiterate a point that was made earlier today but didn't receive quite the attention it should have. An engineering colleague said, "What are you guys talking about up there? What is energy efficiency?" A failure to really define what we are talking about has permeated much of the debate about energy efficiency. The question of "command and control versus market forces" belies some of that confusion. The way I would put it, being an engineer not an economist, is that some of us are talking about economic efficiency, and others are talking about thermal efficiency.

The environmental community is more interested in thermal efficiency than economic efficiency, and in particular the first two laws of thermodynamics. The first law of thermodynamics is, you don't waste energy. You can't create energy, so if you go at it from an energy efficiency point of view, you should try to avoid waste of energy. The second law of thermodynamics, from an energy efficiency point of view, is to try to match the quality of the energy. Amory Lovins made the analogy, something to the effect of, heating water with electricity is like, I forget, doing something with a nuclear explosion: it's just a mismatch of energy sources. Environmentalists are very interested in this because it's the thermal use of energy, and the carbon dioxide or sulfur dioxide that results, that causes environmental damage.

Anybody who runs a business is of course interested, on the micro level, in economic efficiency: how can I minimize my energy input per product of output, and I really don't care about environmental externalities, or any other kinds of externalities, unless I'm handed a regulation from the government. From a macro point of view, policy-makers and economists would try to maximize economic growth within a country, given the energy resources that are available.

I think this confusion between these two definitions of energy efficiency leads to confusion over the objectives: whether you are trying to maximize economic growth, or whether you are trying to minimize environmental externalities. When you look at it that way it becomes somewhat of a false dichotomy. Environment and growth need not be contradictory.

The main point I would like to make today is that, in developing countries, where I spend most of my time, it's really not a debate between command and control and market forces. When one looks at the situation in many developing countries, as a gross generalization, the systems are highly inefficient, not only from an energy or an economic point of view, but from a mechanical, and many other points of view. There are issues revolving around energy systems used inappropriately for political patronage. On the positive side, the distribution of energy at well below market prices is viewed as a social leveler in those places, especially energy for rural use. Energy systems are considered national security items, and in some people's minds it's very difficult to think of energy systems and markets in the same context.

The example of water pumping in the Indian subcontinent came up recently at a World Bank meeting about energy efficiency. Typically farmers in this area are given an energy connection charge, not an energy charge. They pay to hook up their pump to the grid, and when the electricity happens to be on the pump pumps, and when the electricity happens to be off, the pump is off. The energy is essentially free. Somebody at the meeting said we should give them more energy-efficient pumps, and somebody else said that the World Bank wouldn't allow that: we're giving away the energy, and now you want us to give away free pumps too. You're not talking command and control.

This situation in developing countries though, of highly inefficient practice, has led, I think, to total collapse. It is not an exaggeration to say that the power systems in large parts of the developing world are in a state of total collapse. The response to that total collapse has been an appeal by the governments to the private sector to come in and rescue the situation. I remember a newspaper headline in Pakistan in the early 1980s, where the government turned to the private sector and said, "look, we cannot provide the electricity you need, you have to do it on your own." The market is really the only mechanism in the developing countries that is going to work. There is no command and control. The regulatory schemes in developing countries do not exist, so there is no possibility that command and control can work in that regard.

The one regulatory or government action that needs to be taken in most developing countries is that the government needs to get out of the way. It has to stop trying to compete with rational electricity practices and let market mechanisms work. In fact, that is now happening in several dozen developing countries that have revamped their regulations to allow the private sector to put in independent power systems. In my view, that is the single largest positive force for energy efficiency in developing countries. It opens the door for energy service companies and for renewable projects, many of which seem to be smaller-scale and are not very amenable to large-scale utility implementation. So the market really is a good friend of the environmentalists in the developing world. It seems to me that it's a much simpler situation than we have in the developed countries. And there is really very little debate in the developing world.

There is consensus on one item, that economic growth any way you look at it is part of the path to better energy efficiency in the developing world. A second item of confusion I encounter at energy efficiency conferences is that there is very little knowledge about what works in one country, and how that applies to other countries. The World Energy Efficiency Association is defined almost more by what it doesn't do than by what it does do. We have no intention of doing energy efficiency projects, we have no intention of being an advocacy organization. We do intend to spread the word about what works in one country, and what doesn't work in those countries, preferably in the words of the implementers themselves. We are a communications link among energy efficiency organizations and policy-makers. The idea is to spread information about what the best practices might be in energy efficiency, from one country to the next.

The United States Energy Association/Johnson Controls, Inc. Energy Efficiency Forum is held each year at the National Press Club in Washington, DC. This invitation-only event attracts top energy efficiency representatives and government policy-makers.

Mark Ginsberg, Director, Federal Energy Management Program, Department of Energy, highlights the accomplishments of federal agencies' energy efficiency efforts in 1994.

John P. Millhone, Deputy Assistant Secretary, Department of Energy, accepts the Energy Leadership Award in 1994, given for his many years of public service.

(Left to right) Greg Nebel, Johnson Controls; Henri-Claude Bailly, Hagler Bailly; and Doug Decker, Johnson Controls present a check in 1994 for $10,000 to Ed Dalton of the National Energy Foundation to provide scholarships for students to attend the World Energy Council.

Demand and Control vs. Market Pull: This 1994 panel topic drew a variety of responses from (L to R - Howard Geller, American Council for an Energy-Efficient Economy; Mamoru Sueda, Japanese government; Linda Stuntz, former Deputy Secretary of Energy; Hans-Eike von Scholz, European Commission; William O'Keefe, American Petroleum Institute; James Sullivan, World Energy Efficiency Association.

Llewellyn King, founder and publisher of the King Publishing Group, provides a wry examination of the mid-term Clinton energy policy in 1994.

Panelists often became quite animated in making their views known, as did this group discussing utility deregulation in 1994.

Sixth Annual
Energy Efficiency Forum
1995

Sixth Annual Forum – 1995

RETHINKING NATIONAL ENERGY POLICY: CHANGES AND CONTINUITIES

The mid-term elections of 1994, which took place a month after the previous forum, threw a wrinkle into the 1995 affair. The fifth forum optimistically shifted gears from an emphasis on the public sector to a focus on the private sector, buoyed by the conservation and efficiency progress achieved within the federal government since 1991, and emboldened by the policy commitments of the Clinton administration.

In stark contrast, the sweep of Congress by Republicans determined to scrutinize every program and slash federal spending meant a new twist into thinking about energy efficiency. The sixth annual conference theme spoke volumes: "Rethinking National Energy Policy: Changes and Continuities."

New and seasoned members of Congress, as well as representatives of President Clinton's administration and a number of federal agencies, were invited to address the forum. The goals of energy efficiency have not changed, the audience was reassured, only the means used to achieve these goals have come under scrutiny as our political leaders debate the proper size and scope of the federal government.

Thus, the overlying question was, is there any role for the federal government to play in promoting energy efficiency, and if so, what is it? The speeches were impassioned, the debates were vigorous, and the atmosphere was characterized by a sense of urgency. Our selections cover the range of the proceedings and were chosen to convey the mood, underscore the terms of the debate, and clarify what the participants perceived to be at stake.

We begin with a discussion of the federal role in energy efficiency among three members of Congress, an officer of the

Department of Energy, and the vice president of the non-profit organization the Alliance to Save Energy.

One of the purposes of the forum is to call attention to federal projects that are effectively achieving energy goals. This year, three speakers described specific successful programs implemented through the Federal Energy Management Program of DOE. The international aspect of the forum was presented by a representative of the U.S. Agency for International Development.

We present in its entirety a fascinating conversation about the impact of deregulation on the energy efficiency business, conducted by three very knowledgeable players in the private sector and a leader of the environmental organization the Natural Resources Defense Council, which sees itself as a mediator between various public and private interests.

The forum concluded with the presentation of a CD-ROM of environmental technology resources, indication of the high-tech directions being taken both in the invention and marketing of energy-efficient products.

THE VICE PRESIDENT
WASHINGTON

September 19, 1995

Dear Friends:

Welcome to the Sixth Annual Energy Efficiency Forum. I am pleased that representatives from both government and private industry are here today to discuss this very important issue.

The success of our nation's effort to increase energy efficiency and move toward sustainable development depends upon a lasting partnership between the public and private sectors. In reaching this goal, President Clinton believes the federal government, our nation's largest energy user, has a responsibility to lead by example. On March 18, 1994, he signed an aggressive executive order to improve energy efficiency and water conservation in federal buildings nationwide and to increase investments in renewable energy. Through its buying power, the federal government will open new markets for environmental technologies and encourage other industries to use these products and practices.

President Clinton is committed to maintaining U.S. competitiveness in the environmental technology industry. World markets for environmental technologies are projected to expand dramatically in the years ahead as nations pursue the dual objectives of economic growth and environmental protection. In response, the Administration has developed an ambitious policy to take advantage of this fast-growing global market. The plan, *Environmental Technologies Exports: Strategic Framework for U.S. Leadership*, already has helped to promote technological ingenuity and improve environmental protection across this nation.

Your conference is an excellent opportunity to bring together a broad coalition of government officials and industry representatives to generate new ideas to increase energy efficiency for the future. I congratulate you and encourage you to continue your important work.

Sincerely,

Al Gore

Welcome to the Sixth Annual Energy Efficiency Forum. This gathering of government, business and non-profit leaders will provide the opportunity to explore a number of cutting-edge issues in the fields of energy efficiency and conservation.

In the United States, the private sector leads the way in the development of a wide range of energy efficiency technologies that are available to reduce energy consumption throughout our economy. Many industries, commercial properties and homeowners are installing these systems and are benefiting from their cost-effectiveness. Such developments result in the creation of new jobs in the energy efficiency industry, as well as providing environmental improvements.

As we continue to redefine the role of government in the age of information technology, we must continuously challenge both public and private sector leaders to forge new partnerships that will enhance our nation's use of energy resources. Our efforts should be directed toward benefiting the most important energy consumer, the American public.

While everyone may agree to support the goal of energy efficiency, we must work together to identify effective strategies to achieve that goal in the 21st century. We need to ensure that government agencies assist, not interfere with, the private sector in accomplishing these goals.

This forum brings together acknowledged experts in the field to participate in the national debate on how we can best ensure our country's ability to responsibly meet present and future energy demands.

Thank you for your efforts to further this important dialogue in Washington, D.C. and throughout the United States. Best wishes for a successful conference.

Sincerely,

Bob Dole
Senate Majority Leader

Newt Gingrich
Speaker of the House

WHAT ROLE SHOULD THE FEDERAL GOVERNMENT PLAY IN ENERGY EFFICIENCY?

John Ensign, U.S. House of Representatives (R-Nevada).
Jeff Bingaman, U.S. Senate (D-New Mexico).
Robert Walker, U.S. House of Representatives (R-Pennsylvania).
Peter Fox-Penner, Principal Deputy Assistant Secretary, U.S. Department of Energy.
Mark Hopkins, Vice President, The Alliance to Save Energy.

Moderator: This morning, we're going to address what role the federal government plays, or should play, in energy efficiency. We're pleased to have several members of Congress, and representatives of the Department of Energy and the Alliance to Save Energy, to discuss this topic. This is a very timely issue. As you know, President Clinton recently released his national energy policy plan, entitled "Sustainable Energy Strategies: Clean and Secure Energy in a Competitive Economy." You will see that our speakers have widely divergent views. While we may not have a debate between them, we certainly will hear the clash of ideas.

Our first speaker today was elected to Congress in 1994, John Ensign. Congressman Ensign represents the metropolitan area of Las Vegas of the first district of Nevada. But while he represents an urban district, Congressman Ensign is deeply concerned with the entire state of Nevada. Nevada, like New Mexico, is a state that holds a large presence of the Department of Energy, in addition to other agencies. The federal government controls over 87% of the land in Nevada. Given this history, one would not necessarily expect that Congressman Ensign would take on the Feds, yet this is exactly what he has done. Almost from the first day in Congress, Congressman Ensign has actively participated in an effort to dismantle the Department of Energy. He is a leader on the task force created by Speaker Newt Gingrich to pass legislation to eliminate the Department of Energy. Congressman Ensign is also a member of the House Ways and Means Committee, which is a rare honor for a freshman member. In addition, he received special permission from the House Republican Caucus to also serve on the Natural Resources Committee. Clearly, the House leader has a single mission for Congressman Ensign, and that is to make him a leader regardless of the issue of the future of the Department of Energy. He has already taken on this leadership role. We look forward to hearing him and please welcome Congressman John Ensign.

John Ensign: Good morning, you are probably going to hear a little different perspective from some of the other speakers today. I do not have a background in government, and I think if you don't have a government background you do take a different look at things. When you have more of a private sector mentality, the ways that you would address issues sometimes is different than other people whose whole life has been focused around the role of government in society. Being from the private sector, instead of asking how government could help me, I ask "how can government not hurt me?" I think that a lot of people in the private sector come from the same angle. I think the country sent a message last November, and the message was that government is too large, government has tried to micro-manage way too much in Washington and has been very inefficient in doing that, and most people think that they are over-taxed, even though your last speaker said that we were only 25th in the world. That's certainly something that I would like to strive to be number one at, the number one tax rate in the world. The message that they sent was that we need more bang for the buck. We are sending too much money to Washington and we are getting too little for it.

It was mentioned that 87% of the land in my state is controlled by the federal government. People in the West actually have a little different idea of the role of the federal government than people in the East, and that means the members of Congress as well. Many people in the West have quite a bit of distrust for the federal government, and there is a philosophy that we think we can control things in many areas better from the state. Not in all areas. Certainly there is a role for the federal government in lots of areas of our lives. But I believe in a limited role for central government: keep the power closest to where it is usually best, and that is closest to the people.

In 1977 under the Carter administration the Department of Energy was formed in response to the so-called energy crisis. And we all know that the energy crisis was contrived, and that it was certainly not a true energy crisis. The mission for the Department of Energy was to come up with alternative sources for energy, because we are going to be running out of oil here in the next 20, 30, 40, 50 years, depending on who you were listening to at the time. I remember I was just in high school at the time the energy crisis hit, but I remember the predictions of gloom and doom. Certainly I think we would all applaud the effort in that regard. In 1977, 80% of the Department of Energy's budget was dedicated to energy-related functions. In 1995 the yearly budget for the Department of Energy was almost $18 billion, and less than 35% of that now, only $6 billion, was used for energy-related operations. Almost two-thirds of the budget, $10.5 billion, is used for nuclear weapons, nuclear clean-up, and other national security activities. I think that many of you would agree that if you lose your mission in a company, if you lose exactly what you are supposed to do, you get in trouble quickly. Along with many other people, I think the Department of Energy has lost its mission, and simply become a huge morass of a federal bureaucracy.

That's one of the reasons that many of us in Congress are trying to get rid of the Department of Energy. And when we say get rid of the Department of Energy, we don't mean get rid of every function that's within that department. What we are talking about doing is dismantling it as a cabinet-level position, taking the necessary functions it performs for the federal government, and putting those under more efficient federal agencies. The DOE's budget priorities have certainly outgrown their original scope, and they also duplicate functions that are within other federal government agencies. We formed a task force in January 1995, and we worked for months and had dozens of meetings. We broke up DOE into various sections, and had various groups working on various parts to try to dismantle the Department of Energy. We've had meetings with former Secretaries of Energy, including Don Hodel and John Harrington, and one of the things we've tried to do is streamline, consolidate, and eliminate various programs under DOE's jurisdiction, and reduce DOE to a subcabinet-level agency.

I'm here to tell you I'm not sure that we're ever going to dismantle the Department of Energy, it's that much of a disaster. I heard Newt Gingrich the other day give an analogy about how to straighten out Medicare: he said it is kind of like straightening spaghetti, it's almost an impossible task to straighten spaghetti. I think this is a pretty good analogy for straightening out the Department of Energy.

If you don't sometimes pull things out by the roots, they're like weeds: if you just trim them they come back and they come back stronger. I would take exception to the last speaker, who said the Clinton administration is talking a lot about balancing the federal budget. Talking about balancing the budget and actually doing something about it, I think, are two different things. If it was easy to balance the budget up here, people would have done it a long time ago. If it did not take tough choices to balance the budget, people would have done this thing a long time ago. It would not be tough politically if it was easy. So, part of what we are trying to do, one of the reasons we looked at the Department of Energy, is figure out how can we make the federal government more efficient. We don't have all the answers, and I'm not up here trying to say that we do on eliminating this one department. We are also trying to eliminate several other departments within the federal government.

The federal government, by the way, has never eliminated a department. I mean, we've created a lot of them, but it has never eliminated one. Sometimes, just like in a company, certain things lose their reason for existing over a period of time. Sometimes that also happens in the federal government, but they never go away. And that's what we are trying to do by dismantling the Department of Energy. Looking at every single program within the Department of Energy and saying, should it exist? Can the private sector do it better than what the federal government is doing? Should the federal government be in the business of picking winners and losers? A lot of times in an agency like the Department of Energy that's exactly what it does. Instead of the market determining who are the winners and the losers, the federal government is determining that.

Let me just conclude by saying that the Republican Party is deeply committed to research and development at the federal level. But that research and development, we believe, should be done in the basic sciences area that the private sector cannot do. In other words, there are areas that the private sector would have no reason to pursue, and doesn't have the resources to pursue. We believe that is where the role of the federal government comes in. The government can encourage certain areas and certain ways for the private sector to get involved with research and development, but as far as the federal government taking a total role in determining who are going to be the winners and losers in the marketplace, I as an individual, and I think generally Republicans across the board, object.

The future of our country is truly at stake when we are talking about solving our national debt crisis. It isn't only the annual budget deficit that we have to worry about, we also have a $5 trillion national debt. I think that all of us as Americans have to be concerned about the big picture as well, and we all have to be willing to do our part. I don't care whether you are a Republican, a Democrat, an independent, whatever you consider yourself, we all have to be willing to work together for our children's future and for generations beyond that, because right now we are mortgaging their future and it is not right. It is a morally bankrupt philosophy to live by, and it is time to make those changes.

Questions and Answers:

Q: Do we anticipate privatizing the power marketing administrations, like Bonneville or TVA?

Ensign: Actually, that is part of eliminating the DOE. We are going to start with privatizing SEPA, the Southeast Power Administration, and use that to say, "can we do this, can it be done better, is it a better deal for the American taxpayer?" We are going to use that as a study, and then the next two after that will probably be Southwest and WAPA, the Western Area. Bonneville probably never will be privatized, and TVA is not on the table right now. Who knows in the future, if it works really well for some of the other ones. We want to go step by step. It's an incredibly complex process, and we want to make sure the ratepayers get a good deal as well as the taxpayers.

Q: How do we plan to move energy efficiency throughout the nation?

Ensign: The issue is to make the federal government more efficient in all ways, including energy efficient. If you set that out as a goal, you have to plan, you have to strategize, you have to listen to your people on how to make it happen. As a Republican, I would like to applaud a lot of what Al Gore is doing on reinventing government. We can take a lot of things that the private sector has done, come up with different management philosophies, and apply them to energy efficiency as well. But that is really something that is driven from the administration, its not so much driven from the Congress. We basically control the purse strings, not so much the policies within the different agencies; that is more of the administration's role.

Moderator: The next speaker is Senator Jeff Bingaman from New Mexico. The Department of Energy has a large presence in his state. Senator Bingaman has been a leader in energy efficiency. He understands that the purchasing power of energy efficiency is currently about $80 billion a year, and he wants to make sure that we are on the cutting edge, and that the United States has the leading position. He's also very interested in and has been a leader in education, and is a sponsor of the Goals 2000 educational plan. Without any further ado, I would like you to welcome Senator Bingaman, who has been reelected for his third term in the United States Senate and serves on the Armed Services Committee, the Joint Committee on Energy and Natural Resources, and also as the co-chairman of a very important organization, the Alliance to Save Energy. Please welcome Senator Bingaman.

Jeff Bingaman: The topic of your conference is, what role does the federal government play in energy efficiency? In my opinion there is, and should be, a very strong role for the federal government. Report after report indicates that a great gap exists between the current level of energy efficiency that we have realized in this country, and the potential that we have. It is clear that energy efficiency has a great many benefits. Lower consumer energy costs, improvements in the environment, making housing more affordable, creating jobs, increasing our national security.

One reason that the federal government should have a significant role is that many consumers do not know about new energy efficiency technologies, do not know how to assess them. I think federal programs such as EPA's Green Lights, and the Department of Energy's product labeling efforts are very valuable in helping educate the consumer about energy-efficient products. Another reason is that because energy-efficient products often cost more, someone needs to take the lead to encourage consumer use, and the government is an appropriate actor to play that leadership role. That's why model efforts such as the Federal Energy Management Program's pilot programs to implement energy savings, and to move to energy saving performance contracting are so important. I have proposed an amendment to each of the appropriations bills which were considered in the Senate this year, and we adopted the amendments in each case, to set a goal for each federal agency to reduce its energy consumption by 5% in the next fiscal year. The idea is to have ongoing monitoring of federal government agencies and to encourage more of them to enter into contracts to accomplish that reduction. The federal government is the largest consumer of energy in our country. It is certainly correct to point out that we have done a miserable job in the federal government of serving as an example to the private sector or to state and local government. We are very inefficient in our use of energy. It's time the federal government corrected that.

As you are all aware, significant spending reductions are being proposed, and many of the federal programs that have traditionally supported energy efficiency are threatened. I think these cuts are unwise. Clearly we should be about the business of reducing the deficit, and I commend the Republican leadership in Congress for focusing the attention of the country and the Congress on that issue in this first session of 104th Congress. But what we are doing in Congress is not just about reducing the federal deficit. It is also about reordering our priorities, and if the early returns can be used as an indication, this Congress is giving a very low priority to energy efficiency. Across the board, federal programs that support energy efficiency in the Department of Energy, the Environmental Protection Agency, the Agency for International Development – all those programs are taking very large cuts in this year's budget.

Looking toward the future, as pressure increases, we can expect to see even more cuts, unless the energy efficiency community can begin to mount a strategy to indicate to the Congress that there needs to be some priority given to these areas. Two obvious things can be done. First you need the representatives of companies that market energy efficiency products and services to be much more vocal in the debate on Capitol Hill, to demonstrate to the Congress that there are long-term benefits, to say that this is a seven-year effort to balance the budget, it is not a one-year effort, and accordingly we need to have a longer planning horizon than many members of Congress seem to have.

You need to let your representatives and senators know about the need to maintain a federal commitment to energy efficiency, especially the jobs and business opportunities that it represents for individual firms and for the country. One way to do this, of course, is to work through organizations such as the Alliance to Save Energy, and to work on some of the programs that they are trying to promote jointly with industry groups.

Secondly, we very much need to get beyond the Beltway and not just talk to members of Congress with this message. Poll after poll shows that the American people favor the federal government's efforts to improve the environment. Everyone in this room understands the close link between environment and the wasteful use of energy. We have collectively done a very poor job of communicating that message to the American people. So the Alliance is trying to deal with that problem. This fall we have a series of public service announcements geared to average consumers that will appear on television in the next few months. The campaign has three themes: Save Energy . . . Save Money; Save Energy . . . Save Jobs; Save Energy . . . Save the Earth. We believe the campaign can create a new and a positive awareness of the benefits and importance of energy efficiency in the public mind.

I would like to see many of you step up to support this public service awareness campaign. We have very strong supporters in the insulation industry, Blum and Honeywell have made very generous donations to this effort, Johnson Controls also has made a major contribution recently. There are others who believe this is an effective way to impact any future debate about energy efficiency in the country, and I hope they will also participate.

I recall a conversation I had with former Senator Gaylord Nelson some time ago, where he was saying that, when he looked at all the environmental and conservation-related issues that affected the country, our major failing as a country is that we have not been very successful in developing a conservation ethic. My own view is, whether you are a liberal or a conservative, the idea of developing a conservation ethic, which would certainly include within it attention to energy efficiency, is something that you could rally around. I think the federal government can help to rally us around that, and I hope that the efforts that exist today, the useful efforts that the federal government has in this area, can be continued and expanded.

Moderator: I don't know how many of you watch C-SPAN, but our next guest is on C-SPAN quite a bit. He's been there for years as the Republican, in the minority. Since the start of the year, Congressman Walker has served as Chairman of the House Science Committee, where he has used his considerable energy and expertise to rationalize federal support for scientific research. He knows that limited research dollars cannot be wasted on projects that have political support but no scientific merit. Yes, he is no enemy of Science, no proponent of slash and burn. At a time when most Republicans are trying to eliminate cabinet agencies, Congressman Walker has aggressively advocated the creation of a Department of Sciences. He supports the Space Station program, which as you know is very controversial. He also supports research on hydrogen as an alternative fuel. He is not afraid of basic research that may not yield instantaneous results. He embraces the federal government in the area of basic research. A close friend and long political ally of Speaker Gingrich, Congressman Walker is also chairman of the House Budget Committee and chairman of the House Republican Leadership on Institutions and Personnel. Please welcome Robert Walker.

Robert Walker: Thank you very much, I'm delighted to be with you today. I'm pleased particularly to engage in this dialogue over the role of federal government in energy efficiency. The arrival the new Congress and the approach of the 21st century mark an appropriate time to begin rethinking current national energy policy and the role of the federal government. National energy policy as currently espoused by the administration is largely rooted in the past, in particular the 1970s, and it is characterized by a strong belief in the power of the centralized command/control energy bureaucracy, and by a lack of faith in market mechanisms. A few minutes ago when I arrived I heard a spokesman for the administration talking about the fact that this idea of relying on industry and the marketplace is largely one of mythology. Well, I don't think that is true. I think that largely we are leaning to some new definitions of the world, largely spurred on by a revolutionary change in the economy, so we have to begin to adapt national policy for that.

I think that the current energy policy is out of step with the views and beliefs of the majority of the people in Congress, and I believe if you go out and really examine where the American people are at present time, the direction of national energy policy for the past seven years is dramatically out of step with the American people. The current national energy policy is also largely premised on the correction of market imperfection, on the belief that the government is adept at picking winners and losers, and that government spending will continue to grow unchecked. These premises are no longer relevant.

A recent paper by the Director of Regulatory Policies Studies at the American Enterprise Institute succinctly summarizes the situation. In the late '70s and early '80s, according to that paper, energy markets suffered from three serious imperfections. First, energy prices did not reflect true economic or marginal costs of production, due to government regulations at all levels of oil, gas, and electricity markets. Second, energy prices did not reflect externalities; for example, the driver filling his or her tank with gasoline did not in paying for the gasoline cover the cost of the crude, the oil supply disruptions, and the coal burning plants, did not bear the costs of the pollutants it emitted into the atmosphere. Third, reliable information about the relative cost of energy consumption conservation were not widely available, thereby making it difficult for consumers to make intelligent choices.

However, many of these market imperfections have been or are being eliminated. Oil and gas prices are deregulated and electricity is moving in that direction at an ever-increasing pace. Marginal cost pricing has become widespread in the utility industries. Externalities have largely been internalized in energy prices, with the notable exception of those associated with the reliance on foreign oil. And consumers have been given the ability to be well-informed about the trade-off between the use and the conservation of energy. Current energy policy, however, is rooted in the past. It largely ignores the new market reality, a reality which must be acknowledged when rethinking national energy policy.

The Clinton administration's current energy R&D policy also has to be examined in light of what we need to do if we are going to have a real 21st century energy policy. Once again, in regards to R&D, the administration's approach is rooted in the past, a past that believed that all government had to do was to select the appropriate technologies and the appropriate development paths, throw bushels of dollars out the door, and receive a virtual cornucopia of new and economical energy products. Using that approach, the Department of Energy has spent and wasted billions and billions of dollars on boondoggles, such as the rush to develop synthetic fuels, breeder reactors, and gaseous centrifuge enrichment. But all of those things were not the fault of the Energy Department; most of that reflected what was being done and said on Capitol Hill, and many of those projects were pushed on the Energy Department by Capitol Hill. Nevertheless, they were the wrong approach and eventually ended up not producing much of anything after billions of dollars had been spent. We know that the reason this approach did not work is that the government is simply not capable of choosing technologies, of choosing winners and losers in the marketplace. Only the marketplace can choose its winners and losers.

We also know that there are no longer bushels of dollars to throw out the door. The new Congress is absolutely committed to balancing the budget by the year 2002. Funds will be limited largely to those areas where there is an appropriate federal role, namely in basic research. We think that we can fund the kind of basic research this country requires, we can do so within the context of a balanced budget, and we can provide the new knowledge that the new economy will need in order to have a foundation to build upon. But we simply don't have money to use for every pet project that someone brings down the pike.

To give you an example of that, we are deciding to fund university basic research while defunding programs like the Advanced Technology Program (ATP) over at the Department of Commerce. The General Accounting Office did a recent study on the ATP and found that 80% of the money spent by the government and contributed by business on the program, would have been spent by the businesses anyhow. 80% of them said they would have or might have continued to spend the money even if there hadn't been an ATP program. What that says to me is that we can get this spending, we can represent society simply because the economy is changing, we don't need the government involved in spending money that is really counterfeit R&D.

The authorizations reported out of the Science Committee so far this year reflect this new direction. Across the board we are dramatically cutting back, eliminating programs that attempt to pick winners and losers, at the same time we are boosting funding for basic research. When you hear all of the talk coming out of the administration about the fact that we have cut the spending by one-third it is very, very important to take a look at where we cut the spending. Because, in fact, in the area of basic research we have actually increased funding over what the President requested. However, we did indicate that ATP and some of the rest of these programs actually be cut back. But the idea that somehow this nation is abandoning its role in science is just so much nonsense. The fact is that we spend more on R&D than Japan, Germany, the U.K., Italy, and France combined. So we are not abandoning our commitment in this area: we are trying to redefine it, we are trying to re-prioritize it to make certain we spend our money the right way and on the right things.

I want to emphasize that cutting old programs to reduce government spending doesn't mean we are going to turn our backs on the needs of high technology firms. There are direct financial benefits of reducing the deficit, such as lower interest loans. For those people looking to make long-term investments, bringing down the interest rate is a help, a natural economic solution to some of the problems that we face in the entire R&D economy. We are also looking to create a better environment for nurturing innovation, by improving the tax code to make investments in research and development more feasible; by improving the regulatory process so the burden of regulation does not unduly hamper business; by amending liability law so that the threat of lawsuits doesn't stifle new product development; and by looking at our current anti-trust law to further rationalize it in the context of the worldwide competitive marketplace of the 21st century.

We are very, very concerned about our businesses competing in the global marketplace. But we also believe that many of the problems that we are trying to solve in this country relative to R&D and science spending have been created by government, through too much taxation, too much regulation, and too much litigation. In fact, if we would cure those macro economics problems we would go a long way toward making ourselves competitive in the global marketplace. You still need to put the money into basic research, but some of the rest of this would be very much cured if the government would do its job and get out of the way. In the long run, we believe that macro economics solutions will do more to improve U.S. competitiveness than a series of government-administered handouts. This would be good for all business, large or small, high or low tech, manufacturing or service, not just the chosen few businesses selected by the bureaucracy to receive taxpayer debt-financed grants and contracts.

In conclusion, fundamental rethinking of national energy policy is necessary, and this rethinking must take into account the realities of the marketplace, the limits of government, and the fiscal restraints that we now face. Big-government advocates often see this as the worst of times, and the advocates of the old order certainly see this as the worst of times. But those who have new ideas must see this as the best of times, and those who believe that opportunity should be our way to define our economy of the future have to see this as the best of times. This is a historic and exciting age, and what you see happening in Congress is that people are actually identifying the problems, making promises about how we will deal with those problems, putting programs in place, and then meeting our promises. Having said that, do you have any questions?

Q: What mechanism should we employ to get industry to invest in its own research and development?

Walker: First of all, as a part of the tax bill that we will pass in the very near future, we will make the R&D tax credit permanent. The problem with the R&D tax credit as it has been applied in recent years is that no one knew whether or not it was going to be there next year or not, so no one could plan around making investments based upon it. We think we can now lock it in, make it permanent, and that would be a help. We are also working on some ideas to utilize that R&D tax credit in some innovative ways. For example, to allow business to invest in basic research being done at universities, by allowing them to use the R&D tax credit for the money they invest in university research, so that if they invest in facilities, invest in instrumentation, invest even in programs at universities, they would be able to write off those investments against their R&D tax credit.

I also think that when you reduce the burden of taxation to some extent, and particularly when you reduce the capital gains tax and give people a chance to look at a little longer-term view of things, that also frees up some money that can be put into R&D. We'd also like to figure out some ways to provide some incentives for businesses to innovate. For example, we are talking about giving people the opportunity to have tax-free profits for certain products. Taking a look and saying, okay here's an opportunity to make tax-free profits for ten years. So we're looking at some innovative ways to marshal business investment, and not just think that R&D spending has to be a function of government spending, or a partnership with government spending. That's the old way of looking at things, and we'd like to think that we could find some new ways of making people look at their investment dollars.

There are a whole series of things that have happened in the marketplace. For example, when you now look at the production of electric power, most of the regulatory policies and other things that affect its cost have largely been internalized as a part of the overall cost of production of the energy.

I think that government played some role in helping consumers get the kind of information that they want. But I also find that consumers today are far more able to get information than they had been in the past. The whole information revolution allows us to be empowered individuals in ways that are completely different from anything we've known before. The government no longer has to be Big Brother providing the information; there are many, many sources, including industry, and I believe that in light of that we ought to decrease the government's role and increase the responsibility put on those who bring their products to the marketplace.

Q: What is your vision of the role that government should play in promoting energy efficiency?

Walker: I think that one of the major things that government can do is help produce the research and development that gives a variety of energy choices to the country. It's one of the reasons that I've been a strong promoter of the idea of the research into hydrogen. This is one of the areas we have tended to ignore as part of our energy mix. I believe that it has real promise for the next century and we ought to be providing the foundation of knowledge that is needed to decide whether or not that should be a part of our R&D niche. Energy efficiency will result from a variety of competitive energy sources, all of which have promise in the marketplace, and people choosing among those various sources.

If you do what government has done for too long, and decide that there is only one command and control structure, that government will determine how energy efficiency will be done, that government will regulate the marketplace in a way to eliminate competition, the fact is then government will also have to mandate the efficiencies within the monopoly which it created. I believe that the way you reduce the role of government, and therefore the amount that government has to spend, is to increase the amount of competition and thereby assure that efficiencies will be in the marketplace because competition will force the efficiencies to take place.

Q: Do you recognize energy efficiency as an energy source that is different, that carries common goals such as environmental protection, or increasing the competitiveness in the economy, and would you consider a tax credit for another kind of market mechanism to specifically spur energy efficiency?

Walker: Again, I think that the question is whether or not there is a market for energy efficiency. I believe there is. I think there are people who see value in achieving efficiency, whether it be businesses or other kinds of consumers. I know darn well that businesses themselves, in my area, who are large consumers of, for instance, electrical power, want as much efficiency out of that as possible, because it affects their bottom line as well as the bottom line of other people in the area. So, therefore, I think there is going to be a great deal of pressure for those efficiencies to take place. If government, through its tax policy, can promote doing that in a quicker way, that's worth looking at. In my view, if you inspire the right kind of competition in the marketplace you are going to get the efficiencies, because the market will add some demand that they be there.

Q: I have a question that relates to the regulation that governments use for energy efficiency. First you mentioned that the externalities were internalized. I was wondering if in your opinion that meant the environmental externalities associated mainly with fossil fuel use? Also, do you think the regulatory frame that governs environmental damages from fossil fuel burning are in need of sort of a reinvention so that energy efficiency and renewable alternatives to fossil fuel burning could be rewarded through avoiding pipe controls? I think in many cases some regulations will prevent the use of cleaner alternatives.

Walker: First of all, I mentioned specifically that one of the externalities that had not been internalized yet was the use of foreign oil. As to regulations preventing some of the cleanup, I think that is very much the case. The idea, for instance, that we are pushing everybody toward electric cars at a time when there appear to be some solutions on the horizons that do not use the electric car. For example, I recently spent some time in some of the laboratories in Detroit, and saw the new technology that Ford is now experimenting with that basically puts a catalyst on the radiator as a way of cleaning up ozone in the atmosphere. It basically makes the car into a vacuum sweeper for ozone. I saw a test on an actual vehicle, of ozone going in at 163 parts per million and coming out at 26, which is in fact a good deal lower than the macro economics impact might be of a coal-burning plant producing electricity for electric cars. In fact, if we continue to push the regulations toward the idea that you can only use electric vehicles as a way of meeting the pollution test, we may actually prevent ourselves from developing other technology that would allow us to continue to use the present fossil fuel infrastructure, but in a way that actually results in cars being used to clean up the environment. I think we ought to have a whole variety of options out there, and government ought not stand in the way of having those options available to both consumers and to the economy.

Moderator: Our next speaker is not here to throw any bombs at the Department of Energy. One of his jobs is to explain to us where we are going from an energy policy standpoint, at least in the White House. Peter Fox-Penner was with the Department of

Energy and is now serving as senior advisor for technology policy in the Office of Technology Policy in the Office of Science and Technology in the White House. He coordinates the activities of the Committee on Civilian Technology in the Natural Science and Technology Council and provides advice on civilian technology policy. This assignment draws him temporarily away from his position as principal deputy assistant secretary in the U.S. Department of Energy. Prior to coming to the Department of Energy, Dr. Fox-Penner was a vice president at the Charles River Associates, where he practiced economic and management consulting.

Peter Fox-Penner: Thank you, and good morning to all of you. I believe we just heard from Congressman Ensign just told us some of the reasons why a debate on the subject of energy efficiency and its role in creating economic growth, environmental quality, and a good life for the families of this nation really needs to be discussed. This is the issue I'd like to get into today.

The first point that I'd like to make is, that in this administration, our fundamental objective is economic progress – sustainable economic growth. I think we want the same thing Congressman Ensign wants, all the members of Congress want: we want good jobs for our families, we want a clean environment, we want safe streets and good schools. We summarize this via the notion of economic progress versus sustainable economic growth. That is the fundamental objective of this administration.

The question, is how do you get there? We think you get to sustainable long-term economic growth by building a number of elements into the economy. There is widespread agreement in the economics literature, in the long history of political economy, that sustainable growth comes from a positive fiscal climate. You have to manage the economy well, and that is where the size and the scope of the government come in. We need research and development. Every study of economic progress around the world has shown that research and development – technological progress – is the single best predictor of economic performance over the long run. In this country R&D is responsible for between one-third and one-half of all the real GNP growth since World War II. That's a period in which our economy tripled in real terms, per capita income doubled, and we added 40 million jobs. During that period the United States was, without question, the world leader in research and development. Much of it was promoted by the Defense Department, but that is changing and so we need to change, and we recognize that.

Also of obvious importance for economic growth, we must have a well-educated and healthy workhorse. Education and health care are the core of a new buzz word we use today, social capital.

Finally, we need an adequate infrastructure. And by infrastructure we mean not only our highways and bridges, and airports, which we do need and which do need attention in this country, but the environment is itself a natural infrastructure that sustains an economy and that has to be taken care of if we want safe, healthy, productive families and communities.

The reason we need a federal role and a federal presence in energy efficiency is to advance sustainable economic growth. To get there we are going to need research and development, and we're going to need attention to the environment, and it is particularly in those two areas that our energy efficiency agenda contributes. It is an essential part of our research and development programs, it is an essential part of protecting our environment. It is through those two channels that energy efficiency contributes to sustainable economic growth.

I'll just comment briefly on our fiscal climate. I think that all of us are aware that we are blessed with a good fiscal climate, and as the federal budget deficit is coming under control. We have created 7.1 million new jobs since we came to town, and inflation and interest rates are low. We intend to continue this performance, we intend to keep bringing the deficit down, and to continue to create a good fiscal climate. What we need to pay attention to are the other two elements of sustained economic growth, research and development and the environment.

In the 1970s the U.S. was spending the about the same percentage of its GNP on non-defense R & D as Japan and Germany, and our economy was larger. In 1990 we dropped significantly behind Germany and Japan as a percentage of GNP. We have actually dropped to the point where in absolute dollars we are spending less than them, and that is of great concern to us. You just can't grow a business, and you can't grow an economy, unless you invest. But we are on a downward trend in our national R & D investment. The current Congressional budget resolution calls for a 33% drop in non-defense R & D by 2002, in real terms. This is not the way to succeed in the world economy and not the way to build a good economic future.

Now let's talk about the environment for a moment, as energy and the environment are very closely linked. Energy policies and environmental policies are, and should be, linked. As the well-respected British magazine *The Economist* stated quite succinctly: Energy is the single largest environmental impact creator in the world. 70% of the emissions of each of five major pollutants – CO_2, OX, Carbon Monoxide, SO_2, and NO_X – come from fossil fuels.

We have spent a lot of money since the seventies on energy efficiency R & D, on environmental R & D, and on environmental protection in general. Pollution control expenditures run about $47 billion dollars a year in the U.S. today. Maybe there is some room to cut back on these expenditures, but Congress has overdone it with its call for cutbacks.

Let's look at one particular case study and see if it really builds a case for rollback. Let's look at urban air quality. You all know that in 1970 we passed a landmark bipartisan Clean Air Act. What has happened since 1970? The economy has grown by 71% since 1970, the miles we travel in our automobiles have more than doubled, and during that period urban air pollution as a whole has been reduced 24%. That is a wonderful testament to the power of technology, because it is new technologies that brought us this improvement. I will never forget the energy efficiency seminar I went to at MIT, where a distinguished automotive engineer looked at me and said, "do you know that the air coming out of the exhaust pipe of a brand new, well-tuned, present vintage automobile is cleaner than the ambient air in Boston, Massachusetts today?" Technology is a tremendous gain to environmental protection worldwide. We are in the pollution prevention and the environmental protection business, all of us.

On the other hand, one hundred million of us still live in areas where air pollution exceeds one or more of the EPA three-year standards. That is virtually everyone in an urban area. Here in Washington we had 13 ozone alerts this summer. We are still sorting out the science and health threats from particulates, our toxics, and global climate change; so far, these concerns are enough to warrant some effort. There is just no basis for concluding that we should move backwards in environmental protection or cut the funding for environmental R&D.

Let me also talk briefly about energy security. At present the U.S. has an oil trade deficit of $60 billion, which can't be helping our economy, just as a matter of good fiscal management. According to a variety of projections, our trade deficit is expected to increase in oil by another hundred billion dollars by 2020, when we will be 70% dependent on foreign sources. That's not good fiscal management for our economy, and it does raise some security concerns.

It also raises some great opportunities. Let's look at the liquid fuel supply. According to Shell Oil Company's November 1994 forecast document, automobile ownership in the world will double by 2020. That's only 25 years from now. We are talking about more than doubling the number of passenger cars registered in the world in our lifetime. In Shell's own words this scenario, which they think is possible, would require between five new Saudi Arabias and eight new North Seas. It's just common sense that a demand like that on the world oil system raises real economic and geopolitical concerns.

Why are major corporations cutting their R&D? All of you who are in the business world can probably answer this better than I. Our understanding is that tremendous competitive pressures are making them focus more on short-term results, such as process improvements, or marketing, and that squeezes their capital budgets, and encourages them to spend less on R&D. We would like to encourage business to spend more money on R&D. Research and development co-funded in a reasonable and intelligent way with government resources causes business to increase their own research and development. We also think such investment is necessary to continue the progress in environmental protection.

We have brought the deficit down as a percentage of GNP from over 5% to under 2% in just a few years. Under these circumstances, cutting federal R&D investment is like cutting off our nose to spite our face. We think that reduced R&D will lead to a relatively poorer economic future; it will not lead to greater economic growth. It means, relatively speaking, impoverishing ourselves by failing to invest, much like a family that doesn't put away money for its future. If Congressman Ensign and the Congressional leadership think that improving our energy policies is like straightening spaghetti out, I suggest that they reflect on the fact that it takes a lot of effort, probably with very little payoff, to straighten spaghetti out. It is better and more constructive to concentrate on the role of government to facilitate investment, and one of the best ways to do this is to co-fund some R&D with business. That's why we believe a strong federal presence in R&D is necessary and why we oppose Congressional efforts to slash the federal science and technology budget.

Moderator: It is with great pleasure that I introduce our next speaker. Mark Hopkins is no newcomer to energy efficiency. Mark has testified before Congress on a number of occasions. He is vice president of the Alliance to Save Energy, a true leadership organization in Washington. Mark's knowledge of energy issues is like an encyclopedia. He's written testimony. He's probably written a lot of speeches for senators. But more importantly, he has helped to craft legislation in this field, much of which has been enacted into law. He is thoughtful and he is prepared and he has a strong advocacy position. Would you please welcome a real leader, Mark Hopkins.

Mark Hopkins: Yesterday, the United Nations issued its report on global climate change. Nearly a thousand members of the scientific community have now officially said that greenhouse gas emissions are having an effect. The *New York Times* editorialized on this yesterday.

From what a number of our previous speakers said, you wouldn't really know that we continue to waste a billion dollars of the taxpayers money every year on wasted energy use in the federal government. I think you can see what a very tough job the Alliance has in trying to educate policy-makers about energy efficiency and the need to promote it. That's one reason the Alliance has taken the initiative on these public service announcements on television, in radio ads, and print ads, that are going to start running around the country.

The Alliance is a non-profit group, we're bipartisan, in fact, the co-chairs are from both the Republican and Democratic parties. The organization includes leaders of the business community, government, consumer groups, and environmentalists: it's an alliance. Our purpose as an alliance is to bring people together and try to create a consensus to move forward on energy efficiency. We do that by conducting research and demonstration programs, public education, and through policy advocacy. We are funded through some government programs, some foundations, and by about a hundred corporations, many of which market energy-efficient products and services, like Johnson Controls. We are not a trade association, in fact there is no trade association for energy efficiency. We do try to involve the private sector in our activities, policy, and program agenda; in fact that is one of my primary jobs at the Alliance.

Many of the new folks on the Hill have a certain philosophy, an approach or attitude, in which they don't seem to listen a lot. They already know. I happen to be from Lancaster, Pennsylvania. One of our speakers, Bob Walker, was my former Congressman. I've known him for many years. He is very serious. I don't know if any of you picked it up, but I'm not really clear if Bob understands what energy efficiency is. That is one of the jobs that we have to do, to try to educate him about that. But, he is very serious about making budget cuts.

Recently, I was driving my son to a soccer game. One of his friends said, "Mr. Hopkins, we live in a free country," and I said "Yeah." "Well then why do we have to pay our bills?" I tried to explain that while we have to pay our bills, one of the advantages of being a free country is that we get to select what we buy and how we interact. Giving one of Dad's talks about this, interrelated to something in my profession, I mentioned that you could either pay the utility for your energy, or you could spend it on energy efficiency and install insulation and a high-efficiency furnace. My son, who is kind of a wise guy eight year old says "Oh dad, we've already heard that, we learn all about that at school. Energy efficiency, that is how we are going to save the earth." When I went to school we didn't learn about energy efficiency. In fact, being from Lancaster County, people come in to our school and taught us about how a new plant was going to go in and create job opportunities and would be really good for Lancaster County. Well, Three Mile Island created jobs twice: once when it was built and once when it was encased.

Obviously, the federal role for energy efficiency is under tremendous assault. Many of you may not know it, but there is pending legislation right now to eliminate EPA's Green Program. Why? It's an easy target, if you're cutting DOE, and not many people say too much about it. Why stop? Bob Walker's not going to stop. The DOE also is in trouble. From my observation one of the biggest reasons is limited support from, and the lack of a unified voice on the part of the energy efficiency industry, those companies that manufacture and market services and supplies. Some companies, and I could name Johnson Controls, are supportive of the Alliance, involved, doing things, but there are literally thousands of business out there and very few are really active. Business is going to be tremendously affected by what is going on, and we could use your help. We are fighting really hard to preserve what can be preserved relative to energy efficiency in the budget, appropriations, and authorization process.

I think it is pretty clear that the federal role is going to change somewhat over the next few years. Clearly there is going to be less spending, unless something new happens. The cuts, let me point out, are in the general market transformation programs. This is incredibly troubling, because for what seems like a decade, the Alliance tried to point out that we needed, in fact, was to tilt U.S. energy policy toward getting greater adoption in the marketplace. The reality is, and I don't mean to say we are against R & D, but the reality is we have literally hundreds of great technologies that are under-used in the marketplace. The real issue is getting existing technologies implemented, and adopted, and overcoming the number of barriers that make it difficult to adopt technology.

If we are going to educate consumers about energy-efficient technologies and change the marketplace, it is absolutely critical for the private sector to take a stronger stand in supporting energy efficiency policy. As Senator Bingaman pointed out, companies that manufacture these products and services need to look beyond themselves and join together to make energy efficiency work. We need to be as loud as the other energy producing associations. It is very difficult to get the word out on efficiency when other energy associations have the lobbyists promoting increased use of energy. In many ways we do a very good job, we have a number of friends up on the Hill who understand our objectives, who know that efficiency is absolutely key. But as the federal government cuts back on its involvement in energy efficiency, private industry has to step up to the plate and fill the void. To do that the efficiency industry has to be better organized and begin to work with other sectors of their industry: insulation, windows, controls, etc., in raising that tide. At The Alliance, we are moving in this direction, we are developing a series of new ways in which we can specifically involve the private sector in program and policy activities. This is a new effort, it's very specific, it is very difficult, it takes a lot of our time and effort, but we're committed to it.

I want to point out five things we are doing. We recently created an International Advisory Committee to guide our work on international issues, which we are paying increasing attention to. It also includes representatives from U.S. and European energy efficiency companies, from the government, senior people at the World Bank and the European Bank for Reconstruction and Development. The initial project the committee has undertaken developing a plan to promote energy efficiency investment in Russia. Having been there a couple of times, I know it is no easy task, but the idea of creating a thousand-megawatt energy efficiency project in the region is exciting.

Second is the area of exploring energy-efficient products. The Alliance is quickly expanding our programs to support U.S. companies to export energy efficiency products and services abroad. We have developed a new approach called the Real Deal, which involves working with local non-governmental organizations (NGOs) who identify industries and specific companies that want to make improvements, finding out the types of products they want, contacting U.S. firms, and then pulling together an educational trade mission to go down to these countries. As part of the mission, we work with the companies and our staff to do workshops on energy efficiency, but most important, we follow that up with specific site visits out into the field, linking U.S. companies that make products here and companies overseas that want products. We recently completed initial contacts in Monterey, Mexico, and in Portugal. We are planning upcoming visits to South America, the Far East, and Central Europe. We are making energy efficiency happen, not just going around talking to government officials. I want to hold our staff accountable: if you don't come back with a deal, I want to know about it.

We are also at the same time creating new energy-efficiency non-governmental organizations (NGOs) abroad. I think it is one of the most important things we can do in terms of our technology cooperation offers overseas, especially in countries in transition. Because eventually the U.S. will have a much smaller role in these countries, and it is important that we leave behind an infrastructure in which energy efficiency can advance. We want to help NGOs represent the interests of energy consumers, assist them to educate people in environmental efficiency, show them how to raise funds, and how to address government policy. We are about to open up our first overseas office in Kaliningrad, Russia, and we are hiring a talented young Russian that is going to create an Alliance to Save Energy in Kaliningrad as an example of what the Russian people can do for themselves. I want to point out that, in each of these cases, we involve the U.S. energy efficiency industry to become members of these groups and help them with their development.

Another way we are bringing companies into the Alliance is through our new work to promote commercial energy-efficient lighting in the Baltimore-Washington area. This project is geared to helping small- and medium-sized business, and we've created a regional energy efficiency advisory group. The group includes manufacturers, contractors, and the financial community. I'm pleased to say that Riggs National Bank has come to us, and at the beginning of next month they are going to introduce a new energy efficiency loan program in the region. Riggs is going to offer loans specifically to small- and medium-sized businesses to upgrade their lighting and energy efficiency in their facilities. I want to emphasize that this grew out of business involvement.

Finally, I want to announce today that we are in the process of putting together our next committee, which is going to focus on federal energy management. We plan to have representatives of 40 companies that manufacture energy-efficient products and services participate on this committee. While some very good things have been done with federal energy management, I think the business community has a number of very serious frustrations. We've decided to get the business community together, talk about these sorts of issues, decide on some recommendations, and try to do something about it.

Those are some of the things we're doing. The key is that things are changing. I think it is very important, as I said, for the private sector to take more of an active role and step into this breach that is occurring with the beginnings of federal movement away from the energy efficiency table.

Q: Are you thinking more about promoting financing?

Hopkins: One of the members of our board is from Fannie Mae. We've been working with them for a number of years, and with Pacific Gas & Electric, on a new program in which PG&E is offering consumers special loans to upgrade their homes, and in which Fannie Mae is involved in creation of the secondary market. From my perspective, I want to be doing it everywhere, but it's a pilot program and they want to take a look at how it works. Of course, we've been very interested in how we can connect this with home heating systems as a tool to facilitate energy-efficient loans. As the utility industry begins to be more aggressive, they may want to increasingly look for a resource to finance programs for consumers and perhaps secondary markets. It's a very exciting area, and we are even looking to apply the approach in Central Europe.

Q: You mentioned the Alliance's interest in the market transformation programs. I'm curious to know what the reaction of many of the Republicans in Congress is to these programs that were designed to work through market mechanisms, and were designed to operate with markets and promote and use energy efficiency. Do you see, as folks on the Hill become more sophisticated in the differences among the different energy efficiency programs, that Republicans many in fact support this more as time goes on, or do you think that is not likely to be the case?

Hopkins: One of the things we're trying to do is educate them to see it that way. I'm not that hopeful, because so far the response has been "this is corporate welfare." We made a very strong argument that in fact it is not corporate welfare. While a lot of these programs promote energy efficiency, a social good, at some time we create new jobs and new businesses, but we are not doing it for that reason. We are doing it for the social reasons. That's the broad side of the issue that Senator Bingaman mentioned: lowering consumer cost, enhancing national security, improving environmental protection, etc. These are the reasons you want to be making these investments in the marketplace. But there is tremendous pressure up there to find savings wherever they can. Today, energy policy is being set on Capitol Hill.

ENERGY EFFICIENCY STARS IN THE FEDERAL GOVERNMENT

Mark Ginsberg, Director, Federal Energy Management Programs, U.S. Department of Energy.
Paul Fennewald, Maintenance Policies & Programs, U.S. Postal Service.
Victor Petrolati, Chief, In-House Energy Management Branch, U.S. Department of Energy.
Brian Magden, Energy Coordinator, General Services Administration.

Mark Ginsberg: The task I have today is to give you a little bit of an update, and to recognize some of the best and the brightest in the federal government. I want to compliment both Johnson Controls and USEA on having this forum, and giving us the opportunity to share some of the story that is unfolding in the federal sector. It is a quiet success story. One of hard work and one with heroes.

To put it in perspective, many of you probably recognize that the federal government is one of the nation's largest energy users, using some $9 billion a year for all of its energy supplies to power buildings and fleets and industrial processes. What I would like to acknowledge is that it is also the largest energy saver in the world. What other organization do you know that used $2.3 billion less for energy consumption last year than it used in 1985? That is a pretty significant record of saving literally hundreds of millions of dollars each and every year, with the potential of a billion dollars a year if we all do it right.

So my perspective is that there is still opportunity. We know that in all of our 500,000 buildings and our 3.2 billion square feet of floor space there is still significant potential for reducing energy consumption. To respond, we in the federal sector have inaugurated new ways of doing business. One of those is the energy saving performance contracting rule that the Energy Policy Act set up several years ago. Another is reaching out beyond Washington with regional energy action project teams. DOE has teams in Seattle, Denver, Chicago, Atlanta, Philadelphia and Boston.

We also have a new way of doing business with partnerships. You can't do this big job alone, and we know that; we know that it takes more than just our work. We have to partner with industry, with folks in this room, with utilities all around the country. We're partnering with states, with national energy laboratories, and with the federal agencies. The agencies are where it's happening. They are being very successful in taking responsibility for energy efficiency.

The people that you will be meeting have been a part of the group of several dozen who have been nominated by their agencies for the highest recognition accorded to federal energy managers. We want you to have a chance to hear some of the success stories that are unfolding throughout the country, as agencies work hard to implement energy saving performance contracts. Twenty-two are in place and more on the way. Our training classes are over-booked, so you know more are coming. We just awarded Ellis and Liberty Island, and Walter Reed Hospital just acknowledged its solicitation. Randolph Air Force Base has been awarded and is actually installing energy-saving measures. The Presidio National Park is being prepared, Laguna Nigel General Services building is underway.

Equipment is being installed in all of these efforts. I can tell you that at Fort Lewis, for example, 53 hard-working crafts and trades people are helping to reduce the cost of energy in that facility, thus reducing the cost of government, and helping the economy by providing jobs in that community. It was a great delight several weeks ago to initiate a partnership with Public Service of New Mexico and Sandia National Lab, and Kirtland Air Force Base on energy and water. And to be able to say at that time, looking out the corner of my eye at Senator Domenici as I was saying this, there is a billion dollars of savings potential in the federal sector, and seeing him nod when I said "where I come from that is still big money."

The agencies are also working together. Twenty-two of them signed a Federal Procurement Challenge a week ago, agreeing to buy the upper 25% of efficiency products in the marketplace. Now when we collaborate with local and state governments, as we are doing in New York, we are going to have a very large procurement clout that will send a significant message to industry that we in the federal government intend to buy the best products, the most energy-efficient and environmentally sound products, and there is a market for it when you produce it.

Enough of that update. Let's go to the honorees of the day who represent these success stories, who are working hard at three agencies and who have colleagues throughout the federal government who are doing innovative work and developing partnerships.

First is Paul Fennewald. Paul has been on the front lines of the United States Postal Service helping to pioneer shared energy savings against private sector investment to make the postal service more cost-effective, and to deliver the mail more cost-effectively. Please welcome Paul Fennewald.

Paul Fennewald: Good afternoon. I currently serve as the national energy coordinator for the Postal Service. My goal is to make each of the 31,000 postal facilities as energy efficient as possible, and thereby reduce their operating costs. It is quite an honor to be here today and to share with you some of my thoughts about a successful energy conservation program. To me a successful energy conservation program is somewhat like the Holy Grail. You are constantly searching for it, and even perhaps if you have one, you are not sure of it because you're getting so caught up in the searching itself.

In my mind, the three major elements of a successful program are vision, strategy, and projects. Vision leads to strategy or strategies, which leads to projects, and the projects bring about the desired results, which makes everyone in the organization happy. Vision is a beacon that lets people know where you want to go, but to me vision must be personalized so that the organization can relate to it. I said that my goal is to make each of the 31,000 postal facilities as energy efficient as possible. That is somewhat different than if I were to say our goal was to comply with the mandates of the Energy Policy Act. If that were our vision, what would happen throughout our country is postmasters would be running around trying to find out what the Act actually says, interpreting it for themselves, and literally, we would have some chaos within our program. So we have stated our vision succinctly and clearly such that it reaches everyone within the organization.

Once we have our vision in place then we lead to strategy. Strategy is that part of the program that says, "this is how we want to go from point A to point B." Just as there are different modes of transportation to get from A to B, and also different routes to take to get from A to B, there are also different strategies that are acceptable within an energy conservation program. However, in my opinion it is the added features of innovation and flexibility that make these programs successful. These features manifest themselves not so much in the strategies themselves, but rather in how they are implemented.

As an example, one of our strategies is to purchase energy-efficient products, and we recently discovered that by going through the Federal Procurement Task Force we can go directly out to our front-line supervisors, who are the people who establish the requirements for what is purchased for the general organization.

Once we have vision and strategy in place the next element is project. As project, I'm not simply referring to lighting or heating retrofit projects, but rather any activity that supports the strategy. Therefore, a project may involve increased management activities or awareness of how to implement purchasing of energy-efficient products. All of these activities and projects have to emanate from the basic strategy. One of the things that can happen when there is room for flexibility and innovation is that a synergy can start within an organization at different levels.

Another example is that we have employee awareness and training as one of our key strategies. Under this strategy we effectively communicated our proclamation for the Portland Showcase. We were also successful in publicizing the results of our contract with Southern California Edison Invest. When corporate relations heard about the individual publication if you will, they saw fit to form what is called the Green Team, and with the Green Team we now have people working across functional lines, off of a master public relations plan, very much like how most of your businesses operate.

So the three elements of an effective energy program are vision, strategy, and projects. However, to me the key to success is people. Tom Peters in his book *Liberation Management* states that all viewing is personal viewing in the end. I believe if we were to examine all of the conservation programs throughout the country we would find that nothing I said here today is unique. However, the major differences would exist in the people. Effective people lead to effective procedure. Procedure is a tool: processes and procedures exist to be used by people. And I believe that there is not a process or procedure yet invented that is impervious to innovative, creative energy management. Philosophically, we wish to decentralize and integrate energy efficiency throughout our organization. To accomplish this, we must utilize to the maximum extent possible, existing resources, and procedures. People make that happen, and specifically the local facility head for energy management. They're the key, and therefore they must be the focus of the organization's vision. Strategy and project must be developed to support and enhance the efforts of the local facility manager and energy manager. In order to be successful, we must listen and learn from the people involved in the day-to- day.

Ginsberg: Here in Washington we have some of the most complex facilities around the country. We also have a success story of literally saving $100 million annually on the cost of energy use as a result of the in-house energy management program at the Department of Energy. The honoree has the job of managing that program at a time when Congress has said, well congratulations, you have done all of this good work and saved all of this money, but for 1996 there will not be any funding for you. The challenge is great and I'm looking forward to working with Vic in helping to make this next year productive. Our honoree is Vic Petrolati of the In-house Energy Management Program at the Department of Energy.

Vic Petrolati: Thank you Mark. The In-house Energy Management Program has been in existence within the Department of Energy since 1978. Our purpose is to reduce energy consumption and utility costs within the department and to improve energy efficiency in our facilities and operations. Since 1978 we have funded over $300 million worth of retrofit projects. We currently have a return on investment of greater than 25% of that $300 million. As Mark mentioned, we currently are saving over $100 million annually on the retrofit projects that have been implemented through this program.

You heard about the best and the brightest here today. What I am receiving should be termed the longevity award, for working in energy savings, performance contracting, and shared energy savings, probably the longest time in DOE. I actually initiated an effort back in 1984 to look into the legal and procurement barriers to doing energy savings performance contracts within DOE. We identified a number of the barriers, and I worked on the initial drafting of the legislation in 1986. Congress enacted a piece of legislation that has eliminated the procurement barriers and legal barriers for that method of financing contracts within the federal government.

We need to do more Energy Savings Performance Contracts within DOE. Specifically, to continue the energy programs in 1996, since Congress has eliminated funding for this program at DOE. I've been involved in the evaluation and contractor selection for the first two shared energy savings projects that have been accomplished within the Department of Energy, and I've also been involved in organizing our National Performance Review Initiative at DOE for replacing steam plants at DOE sites through energy savings performance contracting. So I've had some good experiences with shared energy savings over the last 10 or 11 years.

I'd like to summarize some of my partial successes, in terms of energy savings performance contracts, and I'd like to provide you with a quick summary of two of our most successful projects. The first project we implemented shared energy savings within the Department of Energy was done at Lawrence-Berkeley Laboratory. This is a very complex facility, and this was a pilot test to determine how easily shared energy savings could be done at many of our complex facilities. A very small percentage of our facilities are actually office facilities: approximately 15% of our square footage is office space. We've got over 120 million square feet within DOE, approximately 15,000 buildings. Over 70% of our building space is either laboratory or production facilities, so a lot of our facilities are very energy intensive.

The building we chose at Lawrence-Berkeley Labs is a combination office/laboratory building. It was consuming approximately $300,000 worth of utilities a year. Basically the project was fairly successful, in that we were able to implement a comprehensive group of energy conservation measures within a laboratory, and payments are being made to the energy service company from both actual and measured energy use reduction within that building. I think a good part of our success is due to the fact that before initiating the request for proposal, we did extensive baseline measurements within that building, so that we had a very good concept of the energy consumption of all the major subcomponents within the building. Lighting energy retrofits were proposed, and electronic ballasts were used. The hot water control valves, which had been a major problem in this building, were all replaced. We had the energy service company put in an energy monitoring and control system for central plant operations. We had a variable-speed drive put on the major supply fan. The low-conductivity-motor water pumps were all replaced with more efficient motors, and the hot water boilers were actually tuned as part of this energy service retrofit.

This is a fifteen-year contract. The construction for all of these initiatives cost approximately $300,000; the savings of the first year of the contract were estimated at about $45,000, actually expanding to about $90,000 in the fifteenth year, and the Department of Energy has actually measured and basically verified the savings within this building. During the first year, actual electrical savings were about 44% greater than anticipated by the contractor; however the thermal savings were about 78% less. Total savings in the buildings wound up to be a little less than 10% from the projected estimated savings. So they did a good job of projecting the savings from this project.

The second most successful project we did is at the Forrestal Building. That's the Department of Energy's headquarters building here in Washington. An extensive energy savings performance contract was done for relighting the entire Forrestal Building. The Forrestal Building is approximately 1.7 million square feet of office space. There are approximately 36,000 lighting fixtures in that building. All of the lighting fixtures were retrofitted: T8 lamps replaced T12 lamps in most of the fixtures. Electronic ballasts, tandem wired, replaced a good number of the magnetic core ballasts. Reflectors were used in many of the fixtures as well, allowing the lamps to be reduced from two to one lamp per fixture. Savings were substantial in this retrofit with DOE able to cut lighting energy consumption by approximately 66% in this building. Lighting was consuming over $600,000 a year and 9 million kilowatt-hours annually, and we reduced it by 6 million kilowatt-hours. The actual dollar savings for this project is estimated at a little over $400,000 a year. We have had a seven-year contract with EUA Cogenics. The first three years of the contract the contractor was taking a good percentage of that savings, approximately 75%; the remaining four years of the contract, the Department is taking most of the savings, 85%. Net present value to the government for this effort is $1.3 million without any investment of retrofit funds. One of the things that made this project so successful was the rebate that we were able to receive from PEPCO for many of the retrofits in this building. PEPCO actually came forward with more than $1 million in rebates for this building, making this building a very successful project for DOE.

Finally, I'd like to point out that we told the contractor what we wanted as far as lighting levels, but we did not tell him how to get the job done. We used a performance requirement rather than a prescriptive requirement in this contract, and this was greatly appreciated by the proposing contractor.

Ginsberg: The last person to be recognized is from our nation's largest landlord, the General Services Administration (GSA). This person I first met when we started to hear rumblings of success from our efforts in our regional support office in New York and the work that they'd been doing with Consolidated Edison. I had the joy of coming up for a ceremony in which Con Ed provided a check for $1.2 million to GSA for the rebate for the project. GSA in New York handles New York, New Jersey, and the Caribbean. It has really taken the lead in looking at projects in a very innovative, very enthusiastic way, and Brian Magden is the person who I credit with helping to pull this together, and to help achieve some of the incredible savings in facilities throughout the New York region. We're delighted to have you meet Brian Magden of GSA, New York.

Brian Magden: Good afternoon. I am currently the regional energy coordinator for the General Services Administration. The section's goal is to conserve energy and water for the effective application of existing technology, good engineering practices, and demand-side management programs without degradation in building performance or occupants' comfort. The branch's energy section is the focal point for all energy initiatives. For example, the energy management section obtained funding for contractual awards worth over $3.9 million for 23 energy conservation projects. These projects were focused on lighting retrofits, replacements, occupancy sensors, solar films, storm windows, LED exit signs, and energy-efficient motors. As a result of this diligence, the Northeast/Caribbean region at the conclusion of FY94 reduced its energy consumption by 21% compared to the amount of energy consumed in FY85. This means that GSA Northeast/Caribbean region successfully met the FY95 goals of 10% and the FY 2000 goal of 20% energy reduction. We are clearly on the way to meeting the 2005 goal of 30% reduction.

In FY94, Con Edison and the GSA entered into negotiations. As a result of these negotiations Con Edison has agreed to perform comprehensive energy audits in ten federal facilities located throughout the five boroughs of New York City. In FY96, Con Edison finally completed the ten audits. These audits will continue to include water conservation opportunities, renewables, and life-cycle cost calculations. Since these audits are at no cost to the U.S. government, the U.S. taxpayer has saved approximately $500,000. This is just another example of a fruitful partnership between the public and the private sectors.

In FY94, GSA opened a new courthouse in lower Manhattan that was built to accommodate the needs of the federal judiciary. This 918,000 gross square foot facility is the largest federal courthouse in the nation. The Federal Building at 290 Broadway is a multi-agency building, approximately 1.2 million gross square feet. The original design submitted by the developers of these two buildings did not include the most energy-efficient systems. For example, the original design called for 40-watt lamps, magnetic ballasts, incandescent lighting in the judge's chamber, non-energy-efficient motors, and non-energy efficient chillers. And of course fluorescent as opposed to LED exit signs. After identifying these problem areas, the GSA energy team met with the developer and Con Ed to begin replacing the proposed system with state of the art energy-efficient systems. As a result of GSA enlisting the aid of Consolidated Edison, a new design incorporating the latest energy-efficient systems was recommended and accepted by the judges' building committee and the developers of both buildings. The new design, unlike the old design, incorporated T8 lamps, solid state ballasts, compact fluorescent, LEDs, energy-efficient motors, variable-speed drives, and chillers that use 25% less energy. As a result of installing energy-efficient equipment, the Federal Building located at 290 Broadway and Foley Square each claim the designation of Energy Efficient Building under Con Edison's Light Energy Building Program.

As a result of these changes, the General Services Administration received a $3.3 million rebate from Con Edison. And finally, the annual cost savings achieved in the Federal Building located at 290 Broadway is $2.1 million, and the annual cost savings achieved at the Foley Square courthouse is $321,000. This translates into a $2.4 million tax savings for American taxpayers.

PROMOTING ENERGY EFFICIENCY ABROAD

Kelly Kammerer, Counselor, U.S. Agency for International Development.

This afternoon, I would like to briefly discuss both the importance of American leadership through its foreign assistance program, and USAID's efforts to promote energy efficiency in developing countries. Recently there has been a great deal of debate about U.S. foreign assistance programs in the press, in the Congress, and by the public. We at USAID welcome the vigorous debate on how we can better target our development efforts to serve U.S. interests. But much of this year's debate has not been constructive, driven instead by those who would have the U.S. turn its back on our leadership role in the international community. To do so would be penny wise and pound foolish. The United States currently spends less than one half of 1% of the federal budget on economic and humanitarian assistance abroad. Of all the major industrialized countries, the U.S. is the least generous donor nation when we look at foreign assistance as a percentage of gross national product.

Yet today's foreign assistance programs are creating the markets of the future for the United States. USAID's programs can help nations develop an enabling environment that creates the opportunity for broad-based, environmentally sound, sustainable economic growth. Our assistance has helped many of today's developing countries to become more prosperous and consequently better customers for U.S. exports. As we phase out assistance to some nations, such as Chile, Argentina, or Thailand for example, we continue to work with poorer countries who will become the big emerging markets of tomorrow.

America could of course, do away with foreign assistance programs altogether, but let us not delude ourselves. Japan, Germany, Britain, France and others will continue to invest in development and will reap the benefits down the road. Japan is already the largest environmental aid donor in the world. The

World Bank and the regional development banks will continue to lend, but without strong U.S. leadership and influence. Do we really want to cut all of our aid programs and walk away from this global market? Trade does not occur without hard work and long-term investment in countries and relationships. Our trade competitors know that, and the businesses in the audience know that. Developing countries sometimes need a helping hand in establishing sound regulatory frameworks, building democratic institutions, improving the efficiency of their industries, and enhancing human capacity. Unlike the trade agencies of the U.S. government, such as the Export-Import Bank and the Department of Commerce, USAID is on hand to provide short-term market development assistance, not just transaction support.

I don't know any CEO who believes you can make a company prosper by focusing exclusively on short-term opportunities. USAID is an important part of the public/private partnership that focuses on both the short- and long-term potential for investment. To give you an example, when I was running the aid program in Katmandu a few years ago, there was a change of government, and for the first time it became a participatory democracy as opposed to an absolute monarchy. There is a lot of potential in Nepal to generate hydroelectricity, but the statutes on the books discouraged foreign investment. With the help of some engineering consultants that we brought in, we were positioned on the ground, we knew who the players were, and we were able to set up a project with the Ministry of Energy to rewrite their laws and regulations covering foreign investment in that sector. Now, for the first time, Nepal allows foreign investment in energy.

To further this partnership, USAID's energy efficiency program benefits U.S. exporters of energy technology, directly by creating demand for investment in specific projects, and indirectly by strengthening policies, institutions, and human capacity in developing countries. USAID has been working for over 15 years to promote energy efficiency in many countries where market forces were only recently introduced. We started working in Thailand, Pakistan, and India when these countries, like most developing countries, had largely public sector dominated economies. Our work has involved institution-building, including creation of the Energy Conservation Center for Thailand and the

National Energy Conservation Center in Pakistan. We have helped showcase U.S. energy-efficient technologies, and worked to develop private sector capacity through training, energy audits, and other energy efficiency practices. We have provided legal assistance, such as drafting the National Energy Conservation Law for Pakistan, and more recently one for Poland. And AID has helped design information and networking programs, such as the Global Energy Efficiency Network, that provide linkages to a variety of energy efficiency resources. USAID has undertaken a wide variety of activities to build an institutional capacity for energy efficiency.

Spurred by the collapse of communism, developing and post-communist countries around the world are now turning to market-based solutions to help solve their problems, and energy efficiency is no exception. Instead of working with governments once reluctant to adopt market reforms, AID is now working in a niche of evolving competitive markets. Governments are embracing open and competitive markets in order to attract foreign capital, and USAID is there to provide technical support. In Central and Eastern Europe energy efficiency was the first USAID program on the ground after the fall of the Iron Curtain. In this case, the ability of energy efficiency to quickly reduce production costs was used as a strategy to help stabilize economies, and begin to turn lumbering parasols into viable private sector businesses. A USAID emergency energy project for Central and Eastern Europe completed 52 energy audits of industrial plants in eight countries in less than two years. At the same time that savings were obtained, most of the paybacks were less than one year, and plant managers were being trained to operate their plants competitively. U.S. equipment was procured and installed, creating an early awareness of both the effectiveness of energy efficiency, and how U.S. products and technical know-how could be used effectively.

Also underway is AID's Eastern Europe Regional Energy Efficiency Project, which focuses on rationalizing energy pricing, privatizing and restructuring the energy sectors, and building a private capacity to provide energy efficiency services. AID is assisting the Polish government to develop a comprehensive demand-side management program based on integrated resource planning as the basis for a one billion dollar energy sector loan. In the Ukraine, the president ordered the privatization and restructuring of the power sector within one year, and we are on hand to assist Ukraine with this ambitious effort. Based on the U.K. model, the Ukraine is planning to incorporate demand-side management into an hourly auction for power. In an interesting twist, this model will have important implications for power industry reforms in the United States, placing us in the unique position of learning from post-communist countries.

USAID is working in other countries in addition to Eastern Europe to help create competitive markets for energy efficiency. In Mexico our 40-year energy efficiency project has developed a pilot program in industrial motor efficiency to create a market for high-efficiency motors. U.S. motor manufacturers have recognized the new market potential. General Electric and BalDor have both offered millions of dollars in low-interest financing to promote the sales of their motors and drives.

USAID is currently working on several exciting new initiatives to address energy efficiency in developing countries, including the Asia Sustainable Energy Initiative, a sustainable energy center in South Africa, and an energy efficiency financing project with similar institutions in the private sector. The Asia Sustainable Energy Initiative is a two-year initiative to develop and finance energy efficiency and renewable energy activities in India, Indonesia, and the Philippines, with an eye toward developing policies and programs to promote integrated resource planning, privatization, and demand-side management. USAID will integrate activities under the initiative with ongoing training and energy efficiency programs at the mission level.

USAID has accomplished a great deal in the fifteen years it has been promoting energy efficiency in developing countries, but the job is not yet finished. Even as countries undertake macro-economics reform, significant barriers still remain. With increasingly limited resources USAID will have to move activities to even higher leverage ratios.

One of the most cost-effective approaches to improving efficiency in the U.S. is the adoption of energy efficiency standards for buildings, appliances, and cars. When done in combination with other market development activities, such as Golden Carrot awards, standards can act as market-pull mechanisms, reducing the risk for manufacturers that their high-efficiency machines will be undercut by inefficient products. The promotion of standards in developing countries has been similarly cost-effective, but it certainly is not a quick fix. Not only do standards need to be developed with particular attention to consistency among countries, but enforcement will be an issue for many years to come.

USAID must also concentrate more on influencing new construction. The energy efficiency market in developing and post-communist countries is currently around $20 billion per year. The vast majority of this market is not retrofit, as it is in the United States, but new construction. With many economies growing at 7% annually or better, within 10 years over half of all industrial plants and commercial buildings will be new. We must develop much better techniques for influencing the investment decisions that go into determining the efficiency of new construction. This should help expand the market for efficient U.S. products.

To summarize, USAID's efforts to expand the market for energy efficiency are wide-ranging. They span the distance from macro-economics reform, energy sector reform, and energy pricing reform, to private sector capacity building, technology demonstration, efficiency financing, and training. With the wave of economic reforms taking place in most developing and post-communist countries, we expect the market for energy efficiency to grow at 10% per year for the foreseeable future. That is in comparison to a projected 4% growth in the industrialized countries.

In addition to our efforts to create free and open markets for efficiency, we are also working to capture the global benefits of efficiency that are not captured in the national economy. Thus, USAID is also supporting programs such as the U.S. Initiative on Joint Implementation, the Countries Studies Program, and the Global Environmental Facility, among others. Ultimately programs such as these will drive the market for energy efficiency even beyond market clearing level. To date, USAID has successfully promoted greater energy efficiency in more than 40 countries worldwide. Our efforts have helped nations develop in an environmentally sustainable manner while creating vast new opportunities for U.S. technologies and services.

Today, the U.S. has a unique opportunity to help developing and post-communist countries establish strong, viable, competitive markets for power. Now is not the time to withdraw U.S. efforts from this global revolution. Now is the time to maintain our leadership, to help set directions that are consistent with both U.S. interests and the sustainable development needs of our friends.

IMPACT OF ELECTRIC UTILITY COMPETITION ON THE ENERGY EFFICIENCY BUSINESS

David Porter, Senior Vice President, Wisconsin Electric Power Company.
Michael Arny, President, Arcadia Resources, Inc.
Paul Addis, President, Louis Dreyfus Electric Power Co.
Ralph Cavanagh, Director, Energy Program, Natural Resources Defense Council.

David Porter: Good afternoon everyone. Clearly, the energy business in the United States is changing and I would like to share with you some of my thoughts, and those of Wisconsin Electric Power Company, on how this business actually will change.

I would like to begin with some current event activity. Nowhere are the changes in the electric business more evident than in the consolidation of the traditional utility company. My company has been a part of that. On May 1st, our parent companies, Wisconsin Energy Corporation and Northern States Power Company, announced a merger of equals to form Primergy Corporation. It would become the tenth largest utility system in the United States, serving five states and having considerable reach in the Upper Midwest area. What is unique about this particular merger is that it takes two companies who have been very much alike, very aggressive in energy efficiency, low-cost producers of the electric commodity, very financially strong companies and puts them together to make an even stronger company. There have been three other major utility mergers announced since May 1. I fully expect there will be more this year and even more next year. There is a trend toward consolidation, and it is a substantial change in the structure of the industry in the United States. It was driven, among other things, by the Energy Policy Act of 1992, which decreed wholesale competition to be the law of the land. Once the genie is out of the bottle, however, I believe that competition will extend to all consumers, retail and wholesale, in the foreseeable future. You can argue on whether it will be in two years, five years, seven years, but it will happen.

What will this new industry look like? What industry will Primergy and companies like it operate in? We envision it to look something like many different boxes and parts of the business. You can take apart the vertically integrated utility and you can determine which parts of that business can remain competitive and which parts should remain a natural monopoly, because they are, in fact, natural monopolies. I'd like to look briefly at each one of these pieces and then talk about what I see as the future of energy efficiency in the context of this kind of vertically disintegrated market.

First of all, the regulated company. Certainly, the wires and pipes will remain a natural monopoly, and will remain regulated by either the federal government or the state government. Clearly, it doesn't make sense to have ten different distribution companies all providing electric service down the back alleys. A natural monopoly service for the delivery of the commodity will remain under regulation.

There will also need to be a coordinating function, the mass buyers and sellers in a new market. This we call a regional power exchange. Recently, there has been a further change in the terminology, and this, in effect, has become what some now call the Independent System Operator. The regional power exchange or independent system operator will continue to be a regulated function. It can do a number of things. It can facilitate the marketing power. Different generators can sell into this regional power exchange. The commodity market, distribution companies, or what I call retail companies, can buy from this exchange, or it can operate the system by brokering bilateral contracts, contracts directly between buyers and sellers. It is the cornerstone of the more competitive electric industry.

We differ from some of our brethren in that we're willing to set up such a regional power exchange with our own systems. Others have not gone that far and are looking to mandatory structure. I know when an order came out in California there was some controversy about the three large utilities in the state being ordered into the regional exchange, the PoolCo, while other utilities could opt in or not. In our view, it could all be done voluntarily, and once it started, you would have everyone joining in. The benefits of the regional power exchange involve open access. All buyers and sellers will have the same access to the power in the stock market. It can facilitate a bilateral contract market and, in this particular model, the owner of the transmission system has no special rights. The owner of the transmission system simply takes service from the regional power exchange, just like any other user in the system. The regional power exchange would have the responsibility to maintain the reliability of the system, to ensure that reactive support is available on peak days and other system reliability is maintained; but the regional power exchange has no interest in the ownership of the particular facility.

The competitive part of the model is of more interest, I think, to those of us involved in energy efficiency. What I've talked about thus far is the regulated part which will continue more or less as is, and all it does is deliver the commodity to the customer. But the rest of it, particularly the generating part of the business, is truly competitive. The growth of the independent power industry has made the large central station power plant virtually obsolete. Co-generation and renewable technologies will provide the bulk of new generation in this country for the immediate future, until the large central station and nuclear and baseload coal plants need to be replaced, beginning 10-to-20 years from now. At that point, we'll see what technology delivers to replace those units. In a competitive business, we're likely to have much more creative solutions than we've had under the regulated business.

But the most exciting action is going to be in the area of retail services, or the energy merchant, and that's what I'd like to spend the rest of my time talking about. We've called this competitive part of the traditional utility the RetailCo, and it encompasses all of those things that are not strictly a natural monopoly. It includes metering services, billing and remittance processing, the whole customer phone center, and the marketing activities. In other words, all of those functions that utilities have traditionally provided, but which do not require a monopoly to provide. Others can provide it just as well. This particular entity, this RetailCo, can buy power from any of the generators that are out there, sell to any of the customers in the universal customer choice regime which we designate. The successful players in this particular business will be those who are able to aggregate energy, both the commodity and energy services, and meet customer needs.

The two business goals of this business are what all energy service companies are dealing with today. Customer attraction, retention and, ultimately, customer satisfaction. The key to this part of the business is that there are many market opportunities, more than simply delivery of these commodities. It's not just the peak kilowatt-hour anymore. A RetailCo can provide a number of things, including customer-owned generation, thermal energy services, hot or cool energy management, electro-technology, power quality, or energy portfolio management. RetailCos could be set up specifically to provide renewable resources. All of the focus groups that we do with our customers indicate that there is a substantial body of customers who are willing to pay significantly more for renewable resources. Customer demand could drive this sector, rather than implementing a regulatory push insisting that utilities develop these kinds of technologies.

I can certainly envision a number of RetailCos operating in what is now the traditional service area of just one utility company. These retail companies will also have the opportunity to develop alliances and affiliations with others. You'll see, and we're already seeing, electric utility companies beginning to partner with telecommunication companies to provide both energy and information services in the home. Marketing can be done together. This will continue: there will be many more alliances and partnerships built up on the strengths of the respective organizations, utilities, and other providers. But, certainly, the key to success in this business is what it is in any competitive business, and that's to get close to your customers, know their energy needs, and be able to meet them better than anyone else. In order to do that, the traditional utility rate, whether it is six cents a kilowatt-hour, or whatever the number is, won't be one size fits all anymore. Pricing will be unbundled, various products and services will be offered to different classes of customers, and the successful participants in this market will be those who can bundle their services to meet customer needs. The key to maximizing market opportunity is to increase the value of the services that you offer, to increase your profit margin. So, that is a view of what the retail company of the future may look like.

I'd like to spend a few minutes talking about what this all means in terms of the energy efficiency services that utility companies have come to provide. Just a few minutes ago at the award ceremony, the federal government was honored for participating in some of the shared savings services, and utility companies were mentioned as having provided rebates as part of their demand-side management in this environment. What does this kind of a structure mean for that kind of a program?

I see the provision of energy efficiency by utility companies evolving into two categories. The majority of it will be value-driven. It will be energy efficiency that is marketed, because it delivers value to a customer who is willing to pay for it through shared savings, or whatever kind of mechanism is there. That will become the bulk of energy efficiency programs for the future. They will pay for themselves.

Another important part of utility energy efficiency programs is that directed toward the low-income consumer, the renter, those areas where the market can not directly reach a particular individual, but energy efficiency is still deemed by either the federal government or the state government to be an important service to be provided by someone. In our view, that can continue to be done, and be done on an individual state level, and it can be done by placing an access fee on the wires charge of that distribution company to pay for that particular service. It may pay for other social benefits the states want to continue. A moratorium on utility shutoffs, for example, could be paid for by an access fee that would be applied to the distribution charge of a particular utility service area. In that way, all customers will pay that social tax. All customers, including the large industrial customer, will pay their social tax, in effect, because that is what it is, for what the state legislature or state regulator determines is something that should be done. I see a place in this competitive future for both kinds of energy efficiency services, in regulated and unregulated environments.

There is no doubt in my mind as we look to the future that the electric utility will be a very different animal. One of the things that utilities will have to get over in the competitive part of the business is the concept of fairness. Traditionally, utilities have tried to set fair rates and to deliver the same service to all customers. But, in a competitive market is ruthless. There will be winners and losers. Some utilities will find they are not very good at this RetailCo game and, so be it, they will go out of business. There will be new entrants in this market other than traditional utilities. Some former utility companies will be able to thrive in this environment, but I think they will be the minority among those existing today. As this industry adjusts to these changes, you will see further mergers and consolidations, as well as spin-offs of parts of business and niche markets being developed by utility companies. It will be a very different environment than what it has been over the last 75 years. It is going to be an exciting time to be in the business.

Competition does not mean the end of energy efficiency programs. In my mind, it simply shifts them to where they ought to be. Those that deliver value will continue to be profitable businesses for their participants.

Mike Arny: I'm here to tell you about the Retail Energy Service Company (RESCO) model for electric industry restructuring. The message I bring is that restructuring the electric utility industry may be the biggest opportunity to increase the delivery of energy efficiency that we've seen in a long time. Many groups seem to very concerned that it will somehow decrease the delivery of energy efficiency. But, I think, if done correctly, restructuring can actually increase the delivery of energy efficiency, and I'll try to explain to you how I think that can happen today.

I'm glad to be following David Porter on the panel here today. His company, Wisconsin Electric, is one of the few electric utility companies in the United States that has taken an unequivocal pro-retail, win-win status. I think that is great. They are trying to move forward, trying to be ready to compete, and taking serious stock of what is necessary to do that.

The RESCO model for utility restructuring is a customer-focused plan. It starts with giving the customer a choice, then it empowers the energy service providers to meet the customer's needs. I think that is different than the approach that is taken in many of the restructuring models. Many of these models start with figuring out what they'll do with generation and what they'll do with transmission. In my mind, those are simply means to the end of providing customer service. I think you get a clearer view of what restructuring requires if you start with the customers and figure out how to give them the maximum choice in obtaining the energy services they want and need.

Furthermore, the retail energy service company model for restructuring is set up to deliver both the individual consumer's objectives and the societal objectives in the energy services industry. It is based on three simple components. First of all, you empower the customer by giving them their choice of an energy service provider and choice of what services they purchase. Secondly, you empower the energy service providers. You give them access to all of the tools and opportunities for serving customer needs. You give them access to the transmission and distribution system, metering services, and everything else on an equal basis, so they are all empowered to compete and deliver customer services. Finally, you collect, through a distribution line charge, funds to meet the societal objectives that are determined to be in the public interest.

This model is a little different than the Wisconsin Electric model in a few key points. First, it doesn't consider that the affiliation between a utility and the wires business with a retail energy services provider in the retail market or a generating company is appropriate in the same markets. Secondly, it takes a very simple approach to the delivery of energy efficiency via the collection of funds for that purpose. It basically says, whatever the state wants to do to increase the delivery of energy efficiency is fine, and it should be done. However, all of the services that are acquired, such as increased energy efficiency, services should be acquired through an open, independently operated competitive process. Those are the two main differences, but I'd say there are more similarities than differences between the RESCO restructuring plan and the Wisconsin Electric plan. I think these differences will decrease over time as the utilities confront what is needed to really be able to compete in competitive markets. They may well find that ultimately divestiture of generation and retail energy services is really the best way for them to achieve competitive position in the marketplace.

When we talk about utility restructuring, I'm always reminded of the former Soviet Union, because I think it is very ironic that by two different paths, the U.S. electric industry and the Soviet economy arrived at the market at about the same time.

One of the key factors of the Soviet economy was using government controlled monopolies to deliver all goods and services. What did that do? That excluded the engine of entrepreneurial drive from their whole economy. In the United States, we use similar government regulated monopoly mechanisms to deliver our electric services and, in some ways, I think the outcome is similar: we have excluded the engine of entrepreneurial drive from the electric industry. Restructuring gives us the chance to correct that, and I think the utilities that are interested in restructuring see this as an opportunity.

Listening to the talk from USAID, I thought maybe we should see if we could get a grant and have USAID help the United States figure out how to use markets in the electric industry. With the current Congress that may not be possible, but given that we'd be spending the money in the U.S. maybe there is some hope.

Freedom of choice is really a fundamental American value. I think a lot of politicians have probably been elected to office for one thing or another, promising freedom of choice to their constituents. And I do have to say that it is one of the American values that is very important to me. I get pretty narrow minded when I am denied my freedom of choice. I think there are a lot of other Americans who feel the same way. In the marketplace, freedom of choice means freedom to choose what services you buy and who you buy them from. Think about all the goods and services that you buy in the economy: it's a very long list. Now think about for how many of those goods and services the government restricts us in choosing our own supplier. It's a very short list: I can only think of electricity. You may be able to add one or two, but it is a very unusual circumstance in the American economy to not be able to choose your supplier. We're used to it in the electric industry, because it has always been that way. But this is an outlier from the normal approach to the U.S. economy. You have a choice of suppliers for almost all goods and services, except who you purchase electricity from.

Since the RESCO plan is a customer approach to restructuring, we should think first about what the customers want. The customers are really interested in purchasing the sorts of things they experience directly, they are not really concerned about how they are produced or delivered. They want lighting service, they want comfort services, heating, cooling and ventilation, they want communication services, they want information services. Electricity and gas aren't on that list. These are merely means to an end for customers. Customers want to get the services they experience directly, they would like to receive them at a lower cost and at the desired level of reliability but they don't much care how they are produced or delivered.

What does society as a whole want? Society wants more environmental protection, increased energy efficiency, increased renewable energy, increased international competitiveness, and increased international security. It's a fairly long list and, given the current economic restraints and political constraints, how are we going to achieve them? Will the old engine of government-controlled monopolies that excludes entrepreneurial incentives from the electric marketplace achieve those objectives? I think you would have to agree that the answer is no, the old engine won't achieve those objectives. No government agency can put up the money to pay for all of the energy efficiency that would be needed to meet those objectives. If the old system is inadequate, what do we need? We need a new engine to drive attainment of these societal objectives, and I think the marketplace can be harnessed to achieve those objectives.

This brings me back to the RESCO plan. The RESCO restructuring plan is based on aligning the interests in the marketplace with the societal objectives to the maximum extent possible, and then setting up a mechanism to capture whatever increment of attainment of those societal objectives isn't delivered by the marketplace. To do this, as I said before, the model empowers the customer, empowers the retail energy services providers, and collects and uses funds to support the attainment of societal objective.

How do we empower the customers? The most important thing is to give them choice of what services they want to buy and who they buy them from. Customer choice is the primary driver for delivery of goods, services, innovation and cost reductions in our economy today. If, at the turn of the century, Henry Ford had been given a government-enforced monopoly over the production of cars, we'd probably all be driving black cars today.

On this note, dealing with the customer, I would like to point out that I think that environmental groups like Natural Resources

Defense Council (NRDC), as well as efficiency and consumer advocacy groups, have been very effective in working with state public utility commissions and utilities in raising the awareness of energy efficiency and renewable energy. The state public utility commission and utilities were used as an available proxy for the customers. But as this restructuring ball has gotten rolling, utilities have pulled out of demand-side management activities because they want to make themselves more competitive. And the environmental, efficiency and consumer advocacy communities are going to have to find a different horse to ride on to get their objectives delivered. I think the challenge to NRDC and the other environmental, efficiency and consumer advocacy organizations is to figure out how to sell their messages directly to the customers. Unless customers can be shown that they will receive real, and what they perceive as direct, benefits from dollars spent on energy efficiency, consumer protection, and environmental protection, these programs won't be sustainable. I think as David Porter indicated earlier. Customers are ready to support the delivery of increased energy efficiency, increased consumer protection, and increased environmental protection, but they need to be sold the program that accomplishes this.

The next question is, how will we empower the retail energy services providers? We need to give them access to all of the tools they need to meet customer needs. That means we need regional transmission tariffs that are easy to use, we need distribution tariffs that are easy to use, and we need one-stop-shopping for electricity transportation service from generator to customer. Think about the transportation system in this country. If every county had a different tariff that you as a trucker had to negotiate in order to drive across the country, it would be very expensive to drive across the country. Right now, in the United States, even that inefficient approach would be a better system than we have today in electricity. We have some utilities with tariffs, we have many utilities without tariffs. Even where utilities have wholesale tariffs in place. Utility boundaries don't relate to the regions across which electric transactions take place. I probably don't need to mention this, as of now retail tariffs do not exist.

I'd like to emphasize that to empower the utilities to ensure that monopolies, regulated-price utilities, cannot use their positions as monopolies to benefit marketing activities in unregulated parts of their business. I think that if we can separate these functions, so that monopolies are monopolies and they are not connected to businesses trying to participate in the competitive market, we can make sure that the playing field in the retail energy services market is level for all participants. Secondly, I think that eventually utilities will find that this separation is actually necessary for them to be able to compete. As long as they have affiliated relationships with monopoly entities, they will, appropriately, have regulators on their back in the marketplace. I think the retail energy service market will be so competitive that they will not be able to effectively compete if they have a regulator on their back.

Finally, there should be a simple system for collecting the funds through an electric system charge for meeting whatever social programs each state wants to institute. States have a variety of mechanisms for doing this, and they should be allowed to continue using those mechanisms to set the level of the access fee, and they should use competitive, independently managed acquisition processes to acquire the services delivered under those programs.

Paul Addis: Let me begin by saying that I don't disagree with much that David and Mike have said. And that's rather interesting, because three or four years ago, as Louis Dreyfus entered this business, things were different. Number one, there were very few conferences, where today there are conferences about every fifteen minutes. David said he has given three presentations in the last 24 hours. The number of conferences indicates that there is something going on in this industry. Second thing is, three years ago I was not invited to speak at any of these conferences. The basic message was that all was well in the utility industry, efficiency was being delivered, and everybody was quite happy with the system as it existed. When I eventually was invited to speak, I got into rather vigorous debates about the need for change in the industry. I think I was seen as representing a threat. It's clear that we are a new competitor that is offering a new service, so I want market access. I make no bones about that, and I'm not going to deny what our goals are. We're here as a profit-making business to deliver a service and great value. Now, the whole debate has shifted. It is no longer a debate about whether competition should occur. It is more of a debate about how it should be done, who should pay the cost of changing from system A to system B, and when should all of this happen.

Somebody over lunch said they thought that there has been more progress made toward a competitive marketplace overseas, as opposed to the U.S. I would disagree. We've seen the deregulation of the industry in places like Latin America and the United Kingdom, but essentially what has happened is taking a state monopoly, the sole provider of the service, and breaking it up into several different companies. It has been a difficult process. Well, in the U.S. we already have a multitude of providers, and I think the debate is so loud and exciting that the way it all breaks up, we'll have a far more competitive market. I think we are already moving toward that relatively quickly.

The other quick observation that I would like to make before I really get rolling here is, "watch out, your industry, the energy efficiency industry, is also about to undergo a sea-change." It will be difficult at times to see what the customer wants, and how to deliver it. I would agree with David that the customer is going to want the energy value created, and he is not going to be interested in how exactly it was created, he is simply going to seek value. There are lots of predators out there who have set up to create that value and to compete with you.

As the two speakers who preceded me said, we are going to have change, and I think it is going to revolutionize the way energy is produced, priced and purchased. Customers may be far more interested in a total solution for their energy, whether it be gas, efficiency, oil products, and the administration services going along with these products. There are a host of new, innovative, market-savvy competitors entering this business. You could run through a list of names, and they would be a who's who of Wall Street, well-known savvy companies that know how to deal in a competitive, rigorous marketplace. They are entering, they sniff something, they are after value. They have done their homework and see what is coming: they see the largest potential energy market in North America.

Banks, for example, have been interested in the independent power producer (IPP) business. They like the security of lending money, in effect, on the back of a regulated utility with a guaranteed long-term contract. Well, that's over, there are no utilities out there buying IPP cogeneration equipment, no matter how efficient the machine is. That's not what they're doing in a marketplace that is so uncertain about what's going to emerge. A utility is making a very prudent business decision not to enter into long-term contracts with a regulatory requirement to get their returns. So the industry is going to change dramatically, and I think the difference between profitable and unprofitable results for these new energy companies will depend upon how well they understand and manage the transformation of this marketplace. Some people will make a lot of money during this transition, others will fall by the wayside.

But most important is the customer. Previously in this arena, my wife and I weren't discerning consumers of our electric bill, we didn't really focus on it, we didn't understand it. This is changing, faster among commercial and industrial customers than among residential consumers, but I think they are waking up as well. Consumers are becoming of the competition. With the information explosion that we all talk about, information is more accessible. Customers understand more about the market they exist within and they are demanding a change. So they are driving it, they create the ultimate opportunity for risks, and the market is huge. Electricity is over $200 billion dollars at retail today in the U.S. Total energy sales are over $300 billion if you exclude transportation and motor vehicles.

My company, Louis Dreyfus, has been around 140 years in the commodity business. We have over $6 billion worth of hard assets: refineries, grain elevators, chips, railcars, all kinds of processing operations. We've been through a number of these transitions and they all look somewhat the same. Different reasons, the pace is different, but there are price disparities when we move from regulated or controlled markets to unregulated markets. Market participants change. You don't know who's on first one day and who is a supplier the next. That is customary in this kind of an early stage. So you should be aware that many industries have gone through this transition.

Why change the system? Because customers believe economic value is equivalent to free market choice, and we find very inefficient mechanisms to get the product to the consumer. That is a basic truism that people believe, and I think we all basically agree with that. The other thing that is so detrimental if you are a utility that wants to hold the line and doesn't want to make change. The regional price disparities in what people pay for electricity, dependent on where you sit, by the luck of the draw, make it difficult to explain that you have an efficient marketplace when you have low prices in the Northwest or the Tennessee Valley Authority (TVA) service territory, where the government is subsidizing the cost of electricity at retail, and then in New York or Long Island prices are two and three times that, or where the wholesale market the price at which utilities buy and sell electricity, versus what people pay at retail, is so dramatically skewed and different. Consumers ask questions and it is very difficult to explain away and say, "Don't worry, we've got an efficient delivery mechanism and everything is okay." So customers want change.

What do they really want? They want the end to a monopolistic system where a utility brings them a rate schedule and says, "Well, you are a class A customer, class B or class C which means your rate is X." They say, "Wait a minute now, I compete in a competitive market, I've got the guy in the Northwest producing the same product and his cost for power is lower. I've got this foreign competitor, I'm competing in a world market, and you are all telling me I've got to go out and export goods if we are going to continue to grow and thrive as a country, so don't tell me about the static rate schedule. I have to compete, I want to make decisions and change things here. Your rate schedule, that doesn't wash."

This new market is going to be very, very different. The product is going to be stylized to what the customer wants. That's what these new predators, these new entrants, know how to do. They know how to style all kinds of products. They can give you fixed price, they can give you indexed price, they can give you a ceiling on price and prices that decline. They can guarantee you a floor on price if you are a producer, they can straddle that, they can give you a ceiling and a floor. They can give you an option on price. They can associate your electricity price with another commodity, with, for example, aluminum. There will be financial profits, there will be prepayments. We have done prepayments, where somebody pays in advance and uses interest rate arbitrage to use their imbedded cost of funds that may be advantageous to get lower prices.

But most important, and I highlight that, customers will provide a total price for energy. The supplier of that price may even monetize the assets at the customers site, he may even take over the chiller, he may take over the boiler in a building, he may monetize them and turn over money to the customer and guarantee a price. We talk about attracting customers and holding on to customers, but a customer isn't going to stay unless value can be created. There are lots of people out there figuring out how to bring total solutions, not a narrow piece of the spectrum, but a total solution. You better look to the big oil companies, you better look to gas companies, you better look to electric companies, you better look to gas companies, you better look to electric companies, you better look to Wall Street, all kinds of companies are ganging up and getting ready, or are ready.

In mature commodity markets, there is liquidity. Positive, I think, there will be a tremendous new customer list available. Customers who will want your product. They'll have a greater motivation, mainly because they will see that there are alternatives. Utilities won't be the sole interface for the customer, or the single most important one. I think there will be new equipment, new technical solutions that will evolve, because the market will drive those solutions. You will have new tools in your toolbox as you approach customers.

What are the negatives? Well, there is going to be new competition which you haven't anticipated, and new skills required. There will be tremendous price volatility and complexity as you deliver total energy solutions: caps, floors, this, that, whatever. You've got to find those skills somewhere. How do you do that? You can try to grow them internally, you can buy them, you can partner for them, there are a multitude of solutions. This is not unique, many industries have gone through that. The problem is, that the management of the company has got to understand what business it is in, and what risks it is taking. That can be a very difficult task, given some of these new supply-side products that are emerging, and dealing with the kind of volatility we've got. There's a cautionary tale here.

In addition, there may be, initially in this landscape, greater savings on the supply side of the meter than on the unit reduction side of the meter, because your business has been around a while, bring value and bringing solutions. The other side of the meter, that supply side, has had less competition so it is more ripe for change, and the customer will make a bigger change initially on that side.

Let me quickly comment on my company. As I said, we are a 140-year-old company. We have a $20+ billion in revenues, over $5 billion in assets, it's a big company, well known in the arenas it competes in. One of the oldest grain companies, meaning rail cars, terminals to store grain, grain silos, in the business. We sniffed an opportunity in energy, but realized that w had significant limitations, and skill sets that we had to fix, if we wanted to be competitive. We wanted to move very quickly, because if you spend too much time trying to learn the new skills and the market opportunity is seized by someone else, you're behind and you missed the whole thing. We decided that we needed a utility partner, but we needed a utility partner in our selfish interest that wouldn't slow us down. We viewed ourselves as entrepreneurial, clever at what we do: we know how to bring efficient products to the customer. We deal with customers, we deal with reliability, our ships arrive on time. There's a lot of money in our business and we do it well.

We needed a partner, but we knew if we found a wrong partner, in effect, a partner who slowed us down, that was death. We went out and visited utilities, but most of them said, "Thank you very much, but we don't like what you're pedaling, and go away." They weren't interested in those kinds of competition, they didn't have the same view of the landscape that we saw coming. Eventually, we found Duke Power. Duke Power is the sixth largest utility in the country. Their service territory spans the industrial heartland from North Carolina through South Carolina, they have 18,000 megawatts of generation. The day we arrived, the senior management said, "Yeah, we see this market changing like crazy, we don't think it's going to stay the same way." They said, "We are a very efficient operator of power plants, we are the only utility in the country that built its own nuclear plant. We can build every kind of power plant and run it efficiently, and we have always attempted to find situations where we could bring our operating skills to the forefront, and we want to own more megawatts. We'd like to be a national utility."

Think about that. There are no national utilities, there are large regional utilities. Tremendous opportunities in this business because there are no national players, there are only large regional players. Well, Duke had an ambition. Their goal was to grow a business nationally. They said, "We need market driven solutions, we need help in figuring out how to price things at market, not at cost, because we think this market is moving toward goods and services being priced relative to the market." So eventually a union was formed. I think that is rather an interesting story, because here are two companies, one with 140 years of experience as a manager of price risk, who has experience at moving commodities around the globe, and here is one of the largest utilities, renowned for its operating capabilities, both thinking that they need each other. This is one of the creative partnerships that was referred to earlier.

So where are we going with all of that? What we do is build up a portfolio of supply. We are moving over a thousand megawatts a day across the U.S. We're talking to retail customers about delivering them value, and maybe most interesting to this group is the cogeneration business. I don't believe we have any merchant plants, plants that have been built solely based upon the market, that the market will determine their pricing. They don't have a guaranteed buyer on the other side of that plant. There have been some accidental merchant plants, where a regulator has said, "I won't let you put your plant in rate base and recover your return." Well, we have what may be the first true merchant energy plant in the U.S.. The Williams Company, the large pipeline gas company, wanted to build a cogenerator at one of their stripping plants where they process gas. It had some compelling economics which would allow them to build electric generating capacity. They did what would be typical. They didn't call us first. They went around in concentric circles, and talked to their local utility, and went a little further away and a little further and said, "We've got this great competitive price, would you like to buy some?" Everybody said that looked really competitive, well I'm having an RFP in two years, I wouldn't mind a little of that peak stuff, about 5% of that looks good, this that or another. Williams was not willing to take on the risk of building a merchant plant They eventually found us. We have guaranteed to take 15 years worth of supply out of that plant at a fixed price over that 15 years. It is a must-take contract. We do not have a buyer on the other side, where we just turn around and found someone else hiding behind the back door. All the risk, in effect, is on us. The Williams Company took our balance sheet, and we took their credit risk. So there are all kinds of new, compelling, interesting opportunities that are emerging, and I think they will bring value to the customer.

The question that I think that Ralph (Cavanagh) and many others may have, and I think it's a very reasonable question, is, will competition harm environmental quality, and can we call providing efficient and affordable power to the less fortunate part of our environmental quality? These are societal goals and objectives. We believe in the free market, but we don't believe in crazed competition solely for competition's sake. Our country has been based on finding new ways to enable the marketplace and societal goals to coexist. The question is striking the right balance. I think it is relatively clear that we can set societal goals and find ways to achieve them, and gain the values of the marketplace. That's environmental quality too. If we're not competitive and people don't have jobs, well, that's part of the environmental quality. I think the marketplace can coexist with these goals, I think the marketplace can bring value. But, the cautionary tale is, the marketplace waits for no company. It won't wait until you decide what you want to do, it moves and it continues to move and it will impact your business aggressively. You may view it as sort of an engineering solution, but I can tell you having met a number of engineering companies who are selling power plants, that you need a market. The marketplace has spoken, so it will impact your business and, for some, it will create great opportunity and great reward.

Ralph Cavanagh: I rather expect that my role is to introduce a constructive note of discord. I will try not to disappoint you, but I think some of you may be surprised at the way I do it. That is, if I was invited to launch the impassioned defense of the integrated monopoly and centralized command and control way of doing business exemplified by the former Soviet Union, I'm going to do my level best to disappoint those expectations. I think I gain a certain credibility from the fact that the environmental community's support for increased competition in the electric industry didn't begin two or three years ago. It stretches back to the campaign that many of us remember, to get Public Utilities Regulatory Policies Act (PURPA) enacted in 1978, to introduce generation competition, and it has stretched seamlessly to the present, where all of the environmental community has been aggressive supporters of the new Open Access Initiative of the Federal Energy Regulative Committee.

I will not appear before you as an opponent of competition. I will insist that it matters very much how the rules are established, how the winners and losers are selected. I will insist that the way that is done remains very much an issue of public policy choice in a variety of forms, both state and federal. There is no single inevitable market, there is no single historical imperative for how to organize or restructure industry. We've got some interesting choices ahead of us. I want to focus on what I think are the most important of those choices, and how to make sure we get them right. I will, in part, agree with my colleagues on their major points, but there are some areas where I think we need to do more. The most important unanswered question is one that was suggested to me just by looking at the organizers and participants of this forum, which exemplifies as well as anything I can imagine the enormously productive partnership on energy efficiency that has grown up over the last decade between the utility industry, government at the state and federal level, and the private sector. That is, as we look at the new industry, can we find a way of delivering efficiency at least as well as we've been able to do with that partnership over the last decade?

A couple of illustrations just to remind you of what you collectively have delivered. The application of energy efficiency standards to buildings, made possible by the partnership, made possible by the common interest in energy efficiency that these institutions have, is the envy of the world today. So are the utility investments in energy efficiency. I don't think this group needs much of a reminder that we didn't send out a bunch of monopolies to stake out the turf on efficiency and exclude others. The role of utilities on energy efficiency has always been as market makers, as market promoters, as investors. They didn't go out and displace the private sector. Where they did their jobs best, they empowered the private sector. California is a good example, where the record from 1990 to 1994 in utility investment in energy efficiency services delivered by the private sector now shows the cumulative net benefit to all parties concerned of $2.2 billion. That is the net value of savings delivered, taking into account all the investments by the utilities and their customers. The figure for New York State over the same period is $1.5 billion. Many other regions can tell a similar story.

The question is, will we be able to emerge with a restructured industry that maintains that kind of congruence of interest, that gives us some hope of delivering that kind of benefit? Indeed, I think we have to aspire to more. I submit that the reason for aspiring more is ever clearer of the headlines of the last few weeks.

Paul gave you the case for electricity as a fungible commodity. In environmental terms, electricity is not just another commodity. The EPA recently summarized just how much electricity matters in environmental terms. In terms of its contributions to annual emissions, the electric industry is responsible for 72% of sulfur dioxide, one-third of nitrogen oxide, and all of you who lived through ozone alerts of the last summer had that brought home to you. One-third of the particulates, likely to be the next great issue of regulatory concern around this country and abroad. One-quarter of the Mercury emitted into the atmosphere, not yet as widely appreciated, but certainly significant feature. Finally, 36% of all of the carbon dioxide produced by economic activity comes from the electric industry. I hasten to say that this never meant for me that this industry should be seen as the principal source of the problem. What those numbers mean to me is that this industry is enormously significant, maybe the most significant part of a potential solution, and we want to be sure as we go forward that those solutions remain possible, and the incentives to deliver them are preserved.

The economic importance of energy efficiency has been brought home to us by the recent record. Something that we have learned collectively in the course of delivering these benefits – and the "we" here is this room, the private sector, the utilities, the governmental partners – what you know, perhaps better than anyone else, are the market barriers to achieving efficiency. You know how hard it is to deliver these benefits. You know the very fast payback thresholds that the market traditionally has imposed. You know the litany of market barriers and market failures that has made your partnership so important in delivering the efficiency benefits of the past decade.

The question you have to ask yourself about all three of the presentations that preceded me is, why should it be easier now? The one thing none of them really gave you was a reason for concluding that those barriers and failures would be less in the market, or that the new institutions they identified would be better equipped to knock them down. There was a great deal of talk in all three presentations about value-driven energy efficiency as the new way we are going to get things done. I won't have an ill word to say of it, and I think there are many sectors in which private businesses, dealing with willing customers, can deliver significant energy savings and, many of you have proved that. But the opportunity for what was referred to as value-driven energy efficiency has been with us since the age of Edison. That is, the opportunity for private companies to go to customers and work with them on mutually profitable ways of reducing energy bills is not new to the restructured industry. It has always been there. We've never blocked people from taking advantage of it, we've never ceded it to a monopoly. The opportunity to do value-driven efficiency has been with us since the electric age dawned. Again, the problem is that in many sectors and, for many customers, there are significant market barriers and market failures to getting the job done. The question has to be, what are we going to do to replace what we found in the past decade for breaking down those barriers, as we move into the restructured industry? I want to agree with Michael Arny that it is, indeed, possible that a restructured industry can do the job better.

But it's going to be difficult for the utilities. They'll be going into that market with enormous potential stranded costs. Their best competitive strategy lies in pushing as much of that stranded cost burden through to the existing customers as fast as possible, by taking advantage of the fact that they own the distribution system as well as generation. The best competitive strategy in this new commodity world lies in signing up as many people as possible to take-or-pay contracts with new price structures, where as much of the price as possible is fixed and as little as possible is variable. In this bleak new commodity world, that's about the best you can do to try to survive, with all of those predators out there whom we've heard about.

Of course, if you believed all of that, then the prospects for the independent energy services provider, or the person trying to deliver value-driven service, look pretty grim. Because remember, we've moved into a world with much more volatile electric prices, we've moved into a world increasingly characterized by take-or-pay contracts, more and more of the bill is fixed, less of the bill is variable. If you are an energy services provider trying to deliver value to customers, you live off the variable portion of the bill, and in this bleak new commodity world, the variable portions of the bill are less and less, and your market opportunities are less and less.

So how do we get out of that? How do we drive toward a richer future? How do we do better? We've had some interesting speculation on that. Michael Arny, for example, talked about what he thought would be some countervailing forces that would push energy efficiency forward in the new world. I want to comment briefly on a couple of the things he mentioned. One countervailing force he identified was greater choice. When customer choice is introduced in this discussion, it turns out to mean two things – one of which is absolutely crucial, one of which I think is somewhat peripheral. I agree with Michael that there is something psychologically satisfying about my ability to decide that I'm not going to send my bill this month to the specific gas and electric company, I'm going to send it instead to Destech, or SMUD (Sacramento Municipal Utility District), or to Louis Dreyfus. Just having that choice lends some psychological richness to my otherwise barren life. I'll acknowledge that, but I don't think that kind of choice by itself does much to change or enrich the kind of electricity service that I'm getting, or to instill a greater appreciation of the values of energy efficiency. The kind of choice that does that, obviously, is the kind that Michael is most interested in offering, which are more interesting options in terms of my equipment and my operational pattern, and I take it that he's set up and has the expertise to help me do that.

But my concern, again, is that that opportunity, the opportunity to deal directly with me as an electricity user who can be persuaded to pay Michael in return for reduced bill, has been with us since the age of Edison. How has it changed in this new world and, in particular, how has it changed if, as we've heard from some of the other panelists, Michael's ability to give me assurance about the savings I'm going to get from his measures over time is rather obviously going to be less?

The other thing that Michael has admonished us about, something that is supposed to draw the greater appreciation of energy efficiency, is something he says is my responsibility, and I think he's right about this. He says that it is the job of people like me, and organizations like mine, to give people more of an appreciation of the environmental consequences of the choice they make in the marketplace. And I want to assure him that, in fact, we absolutely take that challenge seriously, that it is a major part of what we do. But one of the things we have to tell our members is that many of the benefits associated with making the right choices are shared by all society, and cannot be simply brought down to the proprietary level, or converted into proprietary advantage for the individual who makes the choice. Many of the benefits for choosing energy efficiency are in the reduced environmental emissions that deliver benefits that are shared by everybody. Because that value cannot be converted to the individual proprietary advantage of the participating customer, it gets lost in a world dominated solely by what my co-panelists have been calling the value-added transaction.

If you find any force or justice at all in those concerns, what do we do? Does that mean that we simply can't go forward, and my job is to lie down and arrange myself across the rails in front of this on-rushing restructuring engine? I don't think so. I think we can sketch a restructuring model that, in fact, fits what my colleagues have been talking about, and still allows us to get that additional momentum for energy efficiency. Let me close by sketching it for you.

I think the first important thing, and it was implicit in what each of us were talking about, is that the generation and distribution businesses really do need to separate. There is a continuing and increasingly obvious tension and conflict of interest between having an increasingly deregulated generation business, each fighting for the loyalty of every individual customer, on one hand, and having distribution companies that happen to be owned by a few of the generators. That conflict of interest is tougher and tougher to deal with. I thought that David Porter spoke to that very well, and my friendly challenge to him is that, now that progressive elements of the utility industry have acknowledged the inevitable necessity or breaking the link between distribution and generation, somebody needs to go first. I'm sure there are bond analysts present, and I'm sure there are limits on what David can say. I will express my own view that no one would do it better, and my view that whoever does go first will have an enormously productive role in shaping the direction for everyone else. We all seem to think that's going to happen, and I think the crucial thing is to get it going, on a voluntary basis, because I agree with David that it cannot be a matter of government compulsion. I think the word "divestiture" should be stricken from the language, because it has become a synonym for confiscation and government compulsion, and what we ought to be emphasizing now is voluntary efforts to do it right, broadly supported by the other parties. But some of us are getting just a little bit impatient. We've been talking about this for some time. We know a number of companies have been looking at it, but it's not quite happening, and it needs to start.

Once disaggregation starts, I would submit that the next plausible step is for the regulated distribution company – there still will be a regulated distribution company – to assume a role that has been an important part of the existing system, but will now be occupied by a different player, and that is the role of market maker, investor, promoter and, in general, cheerleader for energy efficiency. Michael's company will be delivering the services, but it could use a supportive presence out there, a supportive presence that has the entire nation as its customer base, that is widely trusted, that has a financial incentive for Michael and his colleagues to succeed. The great risk from my perspective is that we restructure and we set up a fundamental conflict between whatever the new version of the utility industry looks like, and the

Michael Arnys, such that the competitive success of one is diametrically opposed to the shareholder welfare of the other. I think we should try to avoid that. I think that the way to avoid that is to have electric distribution companies that are motivated to invest in energy efficiency, break down barriers where they persist, help the Michael Arnys succeed, and get a performance-based reward for doing that. This is also crucial: let's make sure their profits aren't tied to throughout over the distribution system.

I hope my colleagues would agree with me that once we have these independent distribution companies, the last things we want is to have their profits tied to the amount of electricity that moves over the wires. They aren't in the commodity business anymore, their business it to keep people hooked up to electric service. Their profits shouldn't rise and fall with retail electricity throughput. I think that is going to be crucial to the competitive success of the independent energy service providers. The vehicle for doing all of that, for allowing that investment to occur, for allowing the distribution company to perform those function, ought to be a non-bypassable charge on the distribution system. This is what NRDC has called a universal system benefits charge and that has, I think, been endorsed by all the participants on the panel. The only difference among us, I think, is the breadth of the energy efficiency role that we would include. Some of my colleagues appear to think you could limit it just to low-income services. For all of the reasons that I've stated, I think the market barriers in the other sectors will remain significant enough so that there ought to be an investment role for the distribution company through the universal system benefits charge across the whole spectrum of electricity use, and not just the low-income sector.

I think you can manage all of that. You can envision separating generation and distribution, a distribution business with performance-based incentive to invest successfully in energy efficiency, to support and make markets for, and open doors to, an independent energy services industry. And the distribution company's profits are independent of retail electricity volumes. Get all of that and you would, indeed, have an engine for energy efficiency that could plausibly outperform what we have now. Fail to deliver that, and you go back toward the bleaker vision that I sketched at the outset, where we're very much worse off.

The most important message for me, and I think the most hopeful message, is that these choices very much remain to be made. They are not inexorable, that they are not written in the logic of some historical text, they are not the product of some technological imperative. They are classic public policy choices to be made at the federal and state levels, with the assistance of all of you. And to that task, I now remit us all. Thank you.

Questions & Answers:

Porter: Since Ralph gave me a challenge, I would like to respond to it. My company has been ready for 18 months to step into that brave new world that you talk about, and take that step at separating the distribution, the wires and pipes company, from the retail services company, from the generation and from the transmission company. We are prepared to break apart that vertically integrated utility the day after the state and federal legislation removes the regulatory burden from that part of the business that I described as truly competitive. We'll do it that very day.

Arny: I have a little question on top of the last one. David has offered to separate, but that is just an institutional separation, it's not that nasty word, "divestiture." As far as I know, Wisconsin Electric does not propose divesting the generation industry. He said separate, he was careful. Well, let's ask him: I took you to mean that if the federal obstacles were removed, you were prepared to separate the businesses.

Porter: To separate the businesses, but there is not a need to physically divest them from the ownership structure of an overall holding company system.

Arny: I would argue that divestiture is very important, and here is the reason. Affiliated interest transactions are difficult to regulate. One California regulator told me they were impossible to regulate, because they are a long-term relationship. The only way you could possibly regulate them is to be there, present throughout the entire relationship anytime any interaction at all takes place. You can't do that, or at least it's very expensive, and the cost of that regulation is a burden on society. We don't need the affiliated transaction as a society. We'd be better off without them, because life would be simpler. And if those costs are imposed, I would suggest the utilities that foist them on society should certainly bear them. I would secondly offer that, if they choose to bear those costs, they are going to be in a very difficult competitive situation, because they will try to compete with the Louis Dreyfusses of the world who are competing through non-utilities.

Cavanagh: So there has not been competition in that area, but I think that one of the most exciting areas of this whole deregulatory activity is to open up direct, unmitigated competition. I say we should get rid of the word divestiture because of the connotation. I agree with Michael to be clear that it is, in fact, the full set. Ultimately, I hope we will be able to persuade David of this: it is ultimately the full separation of the companies that will be required.

Porter: Public policy should protect competition, not competitors, and that role should not, through regulation, tie the hands of substantial contributors to consumer benefits by precluding them from participating in markets. Markets have a wonderful and brutal reality of creating winners and losers. If there are nearly two hundred investor-owned utilities operating in this country today, I will assure you that the majority, perhaps as many as 80%, will not survive the transition to competition. Let the markets supply the discipline, we don't need regulators to do that.

Cavanagh: I'd like to ask if Wisconsin Electric is willing to bear the cost of the necessary regulation.

Porter: There will be regulation of affiliate transactions in the future, and I will assure you that it is substantially less regulation than exists across the utility industry today.

Questions from the floor:

Cavanagh: The question was, if, in fact, there are significant net benefits from energy efficiency investment above and beyond value of equipment, is there any way to explain that other than the existence of pervasive market barriers and market imperfections? My answer to that is no, *I think you are right, I think that is the basic explanation. I think the burden on anyone who is looking at a new structure of the industry and who wants to make sure we maintain the benefits associate with this partnership I described to find some other way to break those barriers down or describe convincingly why those barrier and imperfections won't exist in the new market. I think that is where so far at least, the other presentations have fallen short. But I would hope that with relative modest amendments, we can find a way clear to solving the problem.

Arny: Again responding to the decision-making barriers, the decision the customer has to make to find the money, find the provider and a lot of complicated things. If the performance contractor came in and said "I'll take care of all your energy service needs and your entire energy budget and reduce your costs," this would remove the barriers for the customer. Changing the regulatory structure so fully competitive, non-monopoly companies can be in both the supply and the demand side business would help overcome those barriers. I don't think it's the whole answer.

Cavanagh: Certainly to the extent that a utility abused the various monopoly powers and advantages that he gets as a regulated company. To compete against independent companies in the marketplace for energy services, that would be a problem. I'm not yet prepared to acknowledge that that is, in fact, an inherent feature to the California system. I think there are some disputes now between the independent energy producers and a couple of the utilities on a couple of the proposals going forward. It is my hope and expectation that those disputes will be settled among the parties involved. The encouragement that I would like to give them is that I've looked at the merits of those disputes, and I'm absolutely confident that the parties involved can work them out.

There have been problems, there has been tension, there has been friction, but I have been impressed by the capacity of the parties to work those things out. Ultimately, what will make this work is we are going to establish the distribution business as the investor, the market maker, the facilitator. That the actual contracts to do the work are going to be won by an independent producer like Michael's. That is where we should be headed, and when utilities want to get into this business, my guess is that they will do it through completely unregulated and independent enterprises.

Q: What will be the impact on efficiency of falling electric rates?

Arny: I think that, first of all, yes, retail competition rates will go down we'll put a freeze on profits. There will be some continued efficiency gains and cross-cutting going on in the industry, there will be some future consolidation efforts to make it more efficient. All of those things will result in winners and losers in that competitive marketplace. Those that can provide low-cost, efficient service will survive, those that can't will die. With regard to elimination of the Department of Energy and the effect on energy efficiency programs, I agree with Paul that the energy efficiency business is going to have to change as well to respond to that new and dramatically new and different marketplace. It will survive and some will adapt and provide those energy efficiency services.

Addis: I guess I would respond in a couple ways. I live in the Northeast and we have lots of trees and lots of storms and our lines go down 2-to-3 times a year, so we already understand what it means to ration electricity, because we lose it without any warning a number of times a year. Now, I am convinced that if you ask my wife, who is in the home more hours than I am, for a price incentive that Michael just described, whether she would take a less reliable service, I think her answer is yes. So again I would agree with a real-life example of what Michael described, and that is ultimate energy efficiency driven by the market. The other thing we have to be very careful about in energy efficiency is that you have to compete with the dollar values that are being created. If the price of electricity was dramatically cheaper relative to the market, those efficiencies might not look so attractive, or they may not have been so real. So I think that the energy efficiency business will have to learn how to deal with the market. It will also have to meet standards set by government agencies in terms of societal goals that we all have to meet, so I think there will have to be a balance created.

Porter: I'm actually more optimistic. I don't think falling electric rates by themselves are going to be a major problem for the efficiency competitors. They've been dealing with that for more than a decade. Real electric rates in this country have declined steadily since 1982; in the industrial sector they are down 25% in real terms over that time. What efficiency providers have learned to do is drop their price, exactly what these gentleman would want them to do. The actual price of the same kilowatt-hour delivered on the California system in 1992 was over three cents a kilowatt-hour, in 1994 it was under two cents a kilowatt-hour. And all of you who have been involved in efficiency investment can attest to similar progress. I don't have any doubt that efficiency providers, given a chance, can out-perform the commodity markets in terms of how rapidly the costs are dropping. My concern again is to continue to make sure that we've got measures in place to deal with the market barriers that deny them the opportunity to do that. I think the questioner is right, that if we don't take any action to address those barriers, we, indeed, will be worse off. The good news is, there are lots of things we can do about that. Remember, we do not have to accept an end to the form of investment that you've been calling rebates, but that all of us have referred to as potentially revivable under the roofwork of universal system benefits charges, or a distribution charge which would allow the utility industry to maintain its role in this area without suffering competitive harm in the process.

ENVIRONMENTAL TECHNOLOGY RESOURCES

John "Jack" Gibbons, Director, White House Office of Science and Technology Policy.
Bill Robertson, Global Environment and Technology Foundation.

Jack Gibbons: You've spent a lot of time today talking about energy efficiency, which has been a very long-term passion of mine. I don't have to convince you that, if we are clever about it, we can do a lot more with a lot less. We can create the goods and services that we need and desire, but at the same time consume less energy and fewer new materials, and certainly with far less environmental impact. The traditional paradigm that somehow economic progress must be accompanied by environmental degradation is just plain wrong, and I think the evidence that has emerged over the past decade or so clearly points in that direction. The opportunities expand faster than the market, and that's what environmental technologies are all about.

I also want to make sure that you understand that the environmental technology industry is itself larger than the U.S. computer industry, larger than plastics, or pharmaceuticals...industries we traditionally thought of as being critical to our economic health. I think we under-appreciated the size as well as the dynamics of this market. The environmental technology industry accounts for over $130 billion in revenues domestically, and the current global market is estimated at around $300 billion. It is also projected that the global market will grow to over $500 billion within the next decade. The industry now employs more than a million Americans in over 40,000 different businesses, and over the next four years it is projected to create and additional 180,000 new high-tech, high-wage jobs. That's an average annual rate of growth more than double that of the U.S. economy as a whole.

The greatest growth in markets for environmental technology will be outside our borders, however, despite the healthy growth within the continental U.S. Foreign markets are growing at an average rate of over 7% a year, nearly 50% faster than the anticipated U.S. growth. The U.S. domestic market for environmental goods and services is by far the largest national market in the world. Consequently, American firms export the lowest percentage of their environmental products abroad. We're far behind, for example, Germany and Japan. Germany's environmental technology industry is a fourth the size of ours, yet they export twice as much as our industry does, and with an annual budget of about $1.5 billion. Japan's environmental technology export is the largest in the world. Japan uses that program to aggressively position its technologies and technology suppliers in foreign markets. On top of that Japan is doubling its research and development budget, which is already about one-third greater per capita than that of the U.S. research and development budget.

Ironically, at a time when competitive pressures on this industry are the greatest, we see Congress withdrawing support from many of the very programs that are designed to ensure continued U.S. leadership and enhance our global competitiveness. There is a play at the Kennedy Center called "Shear Madness," a wonderful play, and I think the title is apt for a play showing in Washington. Another act of sheer madness was the fact that, for the sake of wanting to find a pelt, the Congress has just killed its own Office of Technology Assessment. I guess maybe some people feel that ignorance is the root to salvation, I'm not sure what the logic is. Similarly, the Congress has recently cut FY96 funding for the President's Environmental Technology Initiative by over 70%, which means wiping it out. It's a program that among other things could reduce regulatory compliance cost by $700 million or more throughout our economy. Severe cuts are also proposed by Congress for our programs on solar and renewable energy R&D, as you know, on conservation R&D and on the partnership for a new generation of vehicles. These kinds of draconian cuts not only jeopardize the development of today's environmental technology, but they also raise really serious questions about who's going to develop the next generation of environmentally critical technologies and environmental information, and help capture large, expanding world markets. It could well not be the United States, if in fact Congress goes ahead in the direction they seem to be moving.

Our administration believes that public/private partnerships based on true identification of mutual benefits, on cost sharing, and on peer and merit review, will be a key to leadership in environmental technology areas, as well as other areas where the gains to be had are large, both for the broad public sector and the broad private sector, but where the profits of investing in these technologies are not capturable at the firm level and therefore will not attract private capital investment. There is a lack of understanding by members of Congress working against these things, of how the real world industrial markets operate and how industrial decisions are made. These members assert that even if the government cuts back on supportive research in this area, the private sector will fill in, but we know it is just the reverse. The correlation is that when federal investment moves into an area, private sector investments also move into the area, with perhaps a lag time of about a year or so, and when federal investments move down, the same thing happens with private capital. Congress is operating on some assumptions that I believe are just plain wrong, and are therefore very dangerous to us.

Therefore, we are going to continue to try to remind the Congress that our historic activity of partnering between the public and private sectors for the kind of symbiotic good that can be derived by both is a thing that began when our government began. In the 19th century, it flowered in such areas as agricultural research and the establishment of land grant colleges; it flowered further with the development of manufacturing methods, then of aircraft and aeronautics and aviation, then of electronics and space, and more recently in biotechnology and medicine. All of these things, which now constitute some of our most important industries, had a common base in these kind of partnerships. We are really puzzled about why in the new Congress all of these things seem to be off limits; we are just not quite sure where the logic is, maybe there is no logic. But we in the Clinton administration do believe in these partnerships, and we believe that they will be key in our national leadership and environmental capabilities. The role of government in helping develop and commercialize environmental technology has never been more critical, and, I think, more opportune.

The CD-ROM that we are releasing today is an example of what the government can do. It puts information technologies in the service of environmental technologies. When Vice President Gore released the National Environmental Technology Strategy back in April, following the White House conference last December and meetings across the country, he made a commitment to advance environmental technologies and to support the environmental technology community. To that end we have created a variety of programs that are designed to support technology innovation and commercialization. The programs are built on the recommendations of the industry itself, collected from dozens of those workshops. As a consequence, we launched the rapid commercialization initiative which is designed to help bring new commercial environmental technologies more quickly into the marketplace. Under this program we are trying to help businesses with new environmental technologies to find test sites to test out their techniques, their methods, to verify the performance of those methods, and to acquire the necessary permits to allow their technologies to move more quickly to the market. Under EPA's leadership we will shortly launch a nationwide environmental technology verification program which will help technology developers verify performance and cost data. We believe this program will significantly help our environmental technology businesses to increase their domestic and international market shares.

Four months ago, we created the Inter-Agency Environmental Technology Office, which is a one-stop shop for all federal environmental programs, making government more accessible and transparent, especially for a small business. This CD disk is essentially a one-stop shop. It contains more than 150 full text documents, has guides to associations and environmental technology export opportunities, and 30 videos illustrating innovative applications of environmental technology. It's just extraordinary what you can put on a single disk, in fact it is a little unnerving. You can search all of that information, 680 megabytes of it, in less than three seconds to find what you are interested in using. From technologies for soil decontamination, to export financing programs or cooperative research opportunities. Soon all of the data on this CD will also be available on the Internet, through the Global Network for Environmental Technology, called G-Net, which we've developed to support the environmental technology industry. The CD is a result of all of the federal agencies working together as a virtual agency.

Now I'm going to turn over the presentation to Bill Robertson, who is from the Global Environment and Technology Foundation, and let him surf you through a virtual tour of the federal government's programs on environmental technology.

Bill Robertson: This CD is a road map and a catalogue for getting assistance from the government. It offers the capability to make more profits, so this is a resource for you and I to use to get more from our tax dollars.

The first CD is the Federal Government CD, which is a collection of the capabilities of eleven federal agencies to support the environmental technology industry, and to promote and advance the development and diffusion of environmental technology. It's a tool to help businesses all over the world get the assistance they need from the federal government if they are working in the environment and technology arena.

The second CD is produced by industry, for industry to introduce the capabilities of companies to solve environmental problems worldwide. There are about 1200 companies listed on the CD, and there are feature presentations on nearly 50 companies, that tell their story and present their capabilities for solving environmental problems. These CDs will be distributed worldwide, so I would expect that you would see business opportunities coming out of Indonesia, coming out of Africa, coming out of Europe.

Domestically, we are distributing 50,000 of these CDs. It will be sent to various points of interest around the country where we know that states, federal agencies, and communities need these kinds of services. Environmental technology is a difficult animal, but basically it involves those technologies that address the problems of the human condition, from monitoring evaluations to radiation to waste management, the whole spectrum of those technologies. So if Johnson Controls is involved in the monitoring side, that will open up opportunities not only on the monitoring side, but also with the work that you are doing in waste management and alleviation.

The disk is produced and will become obsolete in probably in a year to a year and a half. It would need to be reproduced. Both the government disk and the industry disk are to be updated with new technology and new capabilities that companies like Johnson Controls have. So we are looking forward to producing a new disk in the near future. While the CD technology is somewhat stable for only a year, the Internet is a very interactive and dynamic technology, so information about Johnson Controls and the other companies and the agencies will be available on the Internet, and can be searched instantly and continuously throughout this period. So we are not concerned that this is going to go stale. We think it is going to become a dynamic tool to help this public/private partnership.

The Energy Efficiency Leadership Award is presented each year to honor certain individuals by recognizing their role in promoting energy efficiency while in public service.

Each forum is professionally videotaped, and participants, such as Johnson Controls' David Matasek, have an opportunity to ask questions on camera.

Senator Jeff Bingaman (D-New Mexico) announces a 1995 public information campaign to increase awareness about energy efficiency.

Rep. Robert Walker (R-Pennsylvania) questions the use of money spent on energy projects in 1995.

Members of a 1995 panel on utility deregulation present several options for utility deregulation. (Left to right) Paul Addis, Louis Dreyfus Electric Power; Michael Arny, Arcadia Resources; David Porter, Wisconsin Electric; and moderator Henri-Claude Bailly, Hagler Bially.

Mark Hopkins from the Alliance to Save Energy analyzes the remarks of several 1995 congressional members about the future of energy efficiency.

1995 was the year of technology, as exemplified in a speech on sustainable development by Jack Gibbons, Director, White House Office of Science and Technology Policy.

Peter Fox-Penner, Principal Deputy Assistant Secretary, Department of Energy, reviews 1995 budget figures for energy efficiency R&D.

Freshman Congressman John Ensign (R-Nevada) challenged many beliefs about the Department of Energy and spending priorities during his 1995 presentation.

In 1995, Kelly Kammerer, Counselor, U.S. Agency for International Development, outlined USAID's involvement in overseas energy efficiency programs.

Rep. Phil Sharp (D-Indiana, right) presents the 1995 Energy Leadership Award to former EPA Administrator William Reilly.

Ralph Cavanagh, Director, Energy Program, Natural Resources Defense Council, illustrates his point during a 1995 discussion on the deregulation of the electric utility industry.

Robert Ichord, Chief, Energy & Infrastructure, U.S. Agency for International Development, (left) receives the 1995 International Energy Leadership Award from former U.S. Rep. Phil Sharp (D-Indiana).

Seventh Annual

Energy Efficiency Forum

1996

Seventh Annual Forum - 1996

ENERGY EFFICIENCY IN CHANGING ENERGY MARKETS

In many ways, the 1996 forum combined the principal concerns of the two previous forums. The theme of the Seventh Annual Forum was "Energy Efficiency in Changing Energy Markets." In addition to changing energy markets, the prospects of international energy efficiency business and issues of deregulation continued to occupy considerable attention.

The morning session was devoted to the question of the appropriate role for the federal government to play in the changing energy marketplace. On Capitol Hill, some of the freshmen Congress members had gotten over their original "slash and burn" philosophy about energy programs and believe that there is merit to energy efficiency. The quandary appeared to be this: if everybody agrees that energy efficiency is good, and if we now can get all 535 members of congress in a bipartisan agreement about an energy efficient ethic, will we shatter the consensus as we move forward to enact it? We reproduce comments from one member of Congress and two officers from the executive branch.

During the international segment Jefferson Seabright of the U.S. Agency for International Development addressed the question of financing energy efficiency investments in the developing world, and the constructive role U.S. businesses can play.

In a departure from the format of the rest of the book, we present three informative presentations from forum participants who wanted to update their thoughts on the impact of deregulation on the efficiency business, including case studies of the American natural gas industry and the British experience with electricity privatization.

Our final selection, which closes out the book, is the address given by former Secretary of Defense Caspar Weinberger upon receiving the International Energy Efficiency Award. Weinberger spoke about energy and national security, striking a reprise that brings us full circle to Congressman Les Aspin's address at the very first forum. Weinberger's speech serves as a reminder that, perhaps, we have not traveled as far on the path toward an energy efficient, environmentally responsible and sustainable economy as we might think in the past decade, and that we still have a long way to go. Most importantly, that energy efficiency is a national security concern.

THE WHITE HOUSE

WASHINGTON

May 15, 1996

Warm greetings to all those gathered in our nation's capital for the seventh annual Energy Efficiency Forum. I am delighted to welcome the many industry leaders who have traveled from across the globe to participate in this important event.

As we prepare to meet the challenges of the next century, restoring and preserving the health and diversity of the world's natural resources is one of our most vital missions. Fostering energy efficiency is an important part of this endeavor, and my 1994 executive order helped the federal government to take the lead in this effort by improving energy efficiency and water conservation in federal buildings and by increasing investments in renewable energy. Through the use of exciting new technologies, we are conserving energy, creating jobs, saving taxpayers' money, and reducing pollution.

As leading members of one of the world's most important industries, you stand at the forefront of the latest trends in a rapidly changing marketplace. By working to encourage energy efficiency and to open new markets and opportunities for growth, you are spurring innovation, improving responsiveness to consumer needs, and helping to conserve the earth's precious energy resources.

I commend the organizers and participants of this year's forum for your efforts to advance these goals. As we work together to achieve responsible environmental stewardship and sustainable economic development, each of you can take pride in knowing that your leadership is helping to create a healthier world for us all.

Best wishes for an enjoyable and productive event.

May 28, 1996

We are pleased to welcome you to the seventh annual Energy Efficiency Forum once again. We hope the forum facilitates a successful exchange of new ideas that assures energy efficiency for the next generation of Americans.

Last year at the forum, we challenged both the public and private sector to forge new partnerships that enhance our nation's use of energy resources. Today, we praise you for your progress to date and renew the challenge for the year ahead.

As you know, energy markets continue to change rapidly. We believe energy efficiency markets can best respond and adapt to market demands without rigid mandates from the federal government. Success can occur only when market forces are allowed to develop freely to respond to America's long-term energy needs. In the spirit of America's conservation tradition, we must continually strive for responsible stewardship of our natural resources in a growing marketplace.

Thank you for your dedicated efforts to lead our nation's accomplishments in energy efficiency. We look forward to hearing about the results of your forum.

Sincerely,

Bob Dole
Senate Majority Leader

Newt Gingrich
Speaker of the House

WHAT POLICIES SHOULD THE FEDERAL GOVERNMENT PURSUE IN CHANGING ENERGY MARKETS?

Scott Klug, U.S. House of Representatives (R-Wisconsin).
Mary Nichols, Assistant Administrator, Air and Radiation, U.S. Environmental Protection Agency.
Christine Ervin, Assistant Secretary, Energy Efficiency and Renewable Energy, U.S. Department of Energy.

Moderator: Please welcome the Honorable Scott Klug, Congressman from Wisconsin's Third District. He is a member of the House Commerce Committee, the committee with jurisdiction over most energy legislation. Congressman Klug also chairs the Speaker's task force on privatization. He strongly favors the sale of a number of energy related assets owned by the federal government, including the power marketing administrations, like the Tennessee Valley Authority, and even the Capitol Power Plant. Congressman Klug believes that assets can be better managed by the private sector, and that their sale can bring billions of dollars of revenue to the federal government.

Scott Klug: Johnson Controls, for those of you who don't know it, is based in my home state of Wisconsin and it is the largest publicly-traded company in the state. So we have a certain warm spot in our hearts for them back home. Let me talk about Wisconsin for a second because I think it raises a broader theme that I want to talk about this morning. Much of the debate in Capitol Hill these days focuses on what level of the federal government is best equipped to handle decisions and whether the federal government can handle them, period, or whether they should be shipped back to the states. So we're in the middle of a political struggle these days over whether welfare should go back to Wisconsin or whether it should be continually run by the federal government out of Washington. The President, a couple of weeks ago in a speech, said nice things about Wisconsin's welfare program, and we've had a vote in the House that now says, Let

Wisconsin do it without waivers. And that's really where much of the debate in the last year in Congress has been. At what level of government is the service best? Should the federal government do it or should state governments do it? I think, and most of my colleagues on the Republican side will say, send it back to the state capitols.

But what's been missing from that discussion about what level of government can best handle a program, is an even more fundamental question: should the federal government even be involved in some of the businesses that it has been involved in for years? I think that's really where the discussions about issues of privatization begin to focus. With a $200 billion deficit this year and a $5 trillion debt, we continue to do some things here in Washington, that, frankly we shouldn't be able to do any more. Period. And more so than anything else, when you look at those privatization issues, many of them cluster and focus right around the energy field. Not energy alone. The air traffic control system of the U.S. was built in the 1950s. It is so out of date we can't even buy the vacuum tubes in the US anymore. We have to buy them from Poland. It's amazing to watch countries like Canada and New Zealand turn around and corporatize their air traffic control system in order to update them and bring them into the 1990s technology. Does the federal government really need to run the air traffic control system, or can it be handled on the private side?

Another interesting story is New Zealand, where they have begun privatizing that country's postal system. And not only have they had a great deal of success, they've had three stamp price decreases in the last 14 months. So there are plenty of stories from around the world that say, we can run these operations much more intelligently and much more efficiently if the government gets out of them.

Now, let's take a look at the Department of Energy. Thirty-percent of the DOE staff, approximately, works running the power marketing administrations. We own and operate 132 hydroelectric dams in the U.S. In 1996, what are we doing in the electric power business? It seems to me, at a time when we're talking about utility deregulation and allowing utilities in Wisconsin to sell power in California and California utilities to sell it to Rhode Island, why do we have these strange islands around the country that are protected from any form of competition? And if you look at the General Accounting Office report, for example, you'll find extraordinarily strong language about the inability of the Tennessee Valley Authority to update and modernize because it feels no competitive pressures whatsoever. And as a taxpayer, and as somebody who is concerned about everybody's taxes in this room, it is extraordinary to think that the TVA has the $30 billion, federally-guaranteed borrowing limit, and it already has $20 1/2 billion in debt. The bottom line is, why does the Pacific Northwest have access to hydroelectric facilities that the rest of the country doesn't have access to, and then those hydroelectric facilities are used, quite frankly, in many situations, as a lure to take business away from the Midwest or Northeast. So, my taxpayers get it twice. They first build it and subsidize it, and then subsequently they underwrite competition to their own electric generating facilities. So let me suggest in the strongest possible terms that I think there is going to come a time in the future, simply from a budgetary perspective, that says, "Are you going to cut Head Start, are you going to cut student loans, or are you going to figure out some way to sell the Southeastern Power Administration?" And I don't think it's very difficult to figure out which way that equation works.

Let's look at another part of the DOE. About 40% of the DOE staff goes to running the DOE labs. But if you look at what the Gavin Commission said, just about a year and a half ago, it made it very clear that a number of those energy labs have outlived their usefulness, or certainly grown beyond their original mission. Anytime we have a bureaucracy, it will mutate and change, essentially, to create a new mission for itself. And the Gavin Commission said, "Wait a minute. There are labs out there that should be consolidated. There are labs that may be closed. There are labs that frankly should have new mission statements altogether." Again, I think the burden is on all of us in Congress to say, "Wait a minute. At a time when we are so far in hock, can we really afford to do this any longer?"

So keep an eye on what I think is in many ways a more fundamental question that has not been discussed with the same kind of passion this year, which is to say, should the government still be in this business, and if the government shouldn't be in this business, how do we get out of it? We've had some success stories this year, but we still have in our possession, for those of you who don't know it, a kind of hodgepodge of oil fields and natural gas fields in California and Wyoming and Colorado, left over from the days when Teddy Roosevelt wanted to enable the US Navy to make the conversion from a wood- and coal-fired fleet to an oil fleet, the Navel Petroleum Reserves. We are going to sell two of those this year and move toward selling some of the additional ones after about another year of study. The GAO studies on the Elk Hills fields in California furnish another good example. Some of them are run by the private sector and some of them are run by the federal government. On the private sector side, there are 12 employees per 100 wells. On the government side, there are 95 employees per 100 wells. What should we be doing in the oil business

Sometimes people will argue, well you have to keep the dams or the oil fields because they make money. Well, if we make money in the federal government running oil fields, then we should be buying them up. And everybody in this room knows that that's not true. We lose money on those operations. And anybody who wins the lottery, given the choice between a million dollars today, cash in hand, or $100,000 over the next ten years, would take the money immediately, simply because everybody in this room understands the time-value of money.

The second issue I want to talk about briefly is, I think we've also got to ask serious questions about how tax policy fits into the picture. It seems to me that if you look at all the debates about tax reform from the last year, there are really two schools of thought. One school of thought advocates moving to a flat tax and making the rates flatter. Make them fair. Take away all the craziness involved in keeping tax returns. We spent about $240 billion in this country keeping complying with the tax code. The tax system is extraordinarily complicated and extraordinarily counterproductive. If you are going to change it though, what do you want to end up with at the end of the day? I think clearly everyone will argue you win with the flat tax because it's simpler and it's easily understood. I think it is simpler to look at a half-page sheet that says, What do you make, figure out 19% of it and send it into the government, without having to fill out all these forms and keep all the documentation right along with it. And there is a certain sense of glee to understand there is not a huge IRS bureaucracy, on the other hand, to send it to. But what you really want to do with a change in the tax system is to fundamentally alter the way the incentives presently work. I think that is a much more compelling argument for a consumption tax.

Now if you ask economists, as Pete Domenici and Sam Nunn did several years ago, what should we do with the US tax system they will say, well, what you want to do is to reward savings and investment. And what you want to do is penalize consumption. If you look at the savings rates across the world, you've all heard these 1000 times, the US saves 2-3% of its income, Germany is in the 9% range, Japan is in the 12-14% range, that's really a fairly clear case to say, we need to reward savings and we need to punish consumption these days. Well, I have to ask my colleagues on the Republican side of the aisle, why are we looking at reducing gasoline taxes in this country? I think for two reasons. First of all, you are going to hear from Cap Weinberger later on, and he's going to point out, as everybody in this room understands, that we continually import about 50% of our oil. And I think that has grave implications for the long-run strategic interest of the US. Secondly, I have an MBA in Finance, and one of the first things you learn in finance theory is the idea of balancing a portfolio. And we've got so much invested today, in oil, and in oil imports and in gasoline, anything that can be done from an economic perspective to discourage that use, and to encourage the development of other fuels or conservation, to me, makes all the economic sense in the world. And I'm not somebody who thinks you should run out, as Ross Perot argued a year ago, and raise gasoline by fifty cents a gallon. I think that would send tremendous shock waves through the economy.

What is it that is going to stimulate the economy the best? Capital gains cuts, arguably . . . putting more money aside in IRAs for student loans, or for medical savings accounts, absolutely . . . because you are sending incentives to have people save rather than have people consume today. But cutting gasoline taxes, I don't think so. I think it sends the wrong signal economically, and I think it sends absolutely the wrong signal in terms of energy efficiency.

The other key component in all of this tends to be scientific research. And the good news last year is that we actually increased the National Institute of Health's budget about 7%, up in real dollars, when many other programs were being cut, money to do medical research and biological research. And while the National Science Foundation budget got reduced slightly, there was more money than ever before actually spent on basic research. We eliminated some of the applied technology programs, which I applaud. I don't think you want to look out there and say, here's a defense contractor that used to make airplanes, now he wants to make buses or taxis. Things don't work that well. The sooner you scrap those kinds of crazy tinkering kinds of programs, the better off you are. Industrial policy doesn't work and it has never worked.

But, at the same time, we need to put more money in one key area and that's renewables. It was interesting to see Dan Schaffer and I last year, two Republicans, essentially lead the fight to restore more money to renewables research in the U.S. Because it seems to me, again, going back to my portfolio model, if you are going to develop a comprehensive and thoughtful energy strategy for this country, then what we should be doing is developing alternatives to importing 50% of our oil. One of the alternatives, clearly, is conservation. That's the business that Johnson Controls is in. The other part of that equation is to figure out a way to make sure the US, in the future, has the ability to switch away from its dependence upon petroleum. Not pie in the sky stuff, but basic, fundamental research on new renewable programs in the country. We are going to have to fight that same kind of battle this time out, because some folks will say, you really don't want to do this, these are applied technology programs for geothermal or applied technology programs for solar. But when you look at the fact that we have been funding coal research in the US since the 1930s, why in last year's budget did coal research go up and renewable technology go down? And why did conservation programs get cut? I'd argue again that if the fundamental mission of the federal government is to do basic research in cutting-edge technology, and to fund the kind of programs the private sector can't do because the payoff may be so distant in the future, then clearly you don't fund coal. You fund renewables and you fund conservation programs, because that's where the greater savings are to be had in the next decade. What started out as kind of a pipe dream with Dan Schaffer and I, actually not only saw us win the amendment last year that restored about $48 million in funding, although at the end of the day we reached an agreement closer to $20 . . . it also saw the creation of a Renewable Energy Caucus, which has nearly close to 100 members at this point.

So, in the big picture, keep your eye on those three trends because I think they are fundamentally important. Privatization: what can we get out of, what energy resources can we turn back to the private sector, where they will be run much more intelligently and thoughtfully. If we are going to do tax reform in this country, what kind of signals do we want to send in terms of investment and savings. And then in terms of consumption, especially in terms of energy use. And finally, if we are going to cut the federal budget, which we obviously need to, what money can we leave behind for basic science research? And in that research, it seems to me, there's got to be a continued federal commitment to both renewables and also to energy conservation. Thanks for the opportunity for me to be with you. I'll be glad to take a few questions.

Q: Concerning the question of selling off the hydroelectric power utilities, it seems like a good idea in some respects, but what do you do with the concept of who does the water belong to? You can sell the dam, but the water is falling over a large area and if some guy built a dam five miles downstream, this dam would no longer work. It's very tied in with water conservation issues. We've always viewed water supply as pretty much a public problem and not something that could be easily privatized, especially in areas where water is scarce.

Klug: Let me walk you through some ideas about the power marketing administrations. There is a whole series of them. We actually just sold the Alaska Public Power Marketing Administration, although that's just a handful of dams that everybody wanted Alaska to run on its own. There's the Southeastern Power Marketing Administration, Southwest, the Western Administration, Bonneville, and TVA, kind of walking your way up the food chain. I don't necessary know that we know how you handle each of those transactions. In some cases, you may want to literally sell the dam. In some cases, you may just want to sell the power from the dams. In some cases, you may want the federal government to retain responsibility for the dams and dam management because that is appropriate from the point of view of the larger water resource issues. In some cases, you may not want them at all, you may just sell them lock, stock, and barrel, and say you have an obligation to hold onto this and take care of it. And obviously, if you don't exercise good stewardship you don't have the power to sell so you lose money on the deal.

Let me say again, though, from a conservative economic analysis, I'd argue in many cases, the federal government's continued involvement in the power marketing administrations, particularly out West, has in many ways undervalued the cost of that water. I'd argue that essentially some of the development and some of the costs you've seen out West don't actually reflect the true price of running the power marketing administration, and don't reflect the true price of the federal government essentially underwriting some of the subsidies in terms of water being redistributed throughout the West. So I think if you sell those you get market forces in place, which I think may make folks more aware of the true price and make them more likely to conserve in the future, rather than to squander it the other way around. I'd argue by artificially creating these islands which have no competition, which kind of have captive markets, you've sent all the wrong economic signals in order to hold onto and price scarce resources. So my argument is that we all will be better served by privatizing it rather than having the government do it.

Q: Members of your party, the majority party, have been rethinking their views on the environment. How do you respond to the argument that energy efficiency is now part of the environmental agenda as far as members of Congress are concerned, or do they still not quite get the connection?

Klug: The question is, since some of the early environmental battles last year, have the environmental opinions of some Republicans changed in terms of being more pro-renewable during this election season? I think if you go back, there is a reason for a lot of us to be skeptical about the government's involvement in renewables, because it seems to be, in many cases, what we've tried to do was force technology before it was available. If you want to look at a perfect example of industrial policy gone crazy, look at the Synthetic Fuels Corp. in the late 1970s. It was an absolute disaster, we poured a lot of money down the hole and got nothing as a result. So I think on the Republican side of the aisle there is a real skepticism about government's ability to say, here's a new technology and we are going to stick all kinds of money into it to get this technology down the road. I think often what you see whenever the government gets involved in those kinds of technology issues is, we essentially dedicate ourselves to a project, whether it's the SST which we tried to do in the 70s, the Supersonic Transport, or a lot of other stuff, and the technology never gets developed. So I think going into the session there was a commitment by some people on the Science committee and Energy committee that said, we're going to get out of a lot of these programs, which look to us a lot like the Synthetic Fuels Corp. We tried to say, okay, maybe you don't want to put the same kind of money into technology application that you thought, but for God sakes, then keep the basic money in solar research, and keep the money in hydro and geothermal to do basic renewables research. Let the private sector make that conversion. But because you may not like the fact that some of the money is now directed at solar applications, don't wipe out everything behind it.

And I think that fits fairly coherently into the argument I made before. One of the missions of the federal government should be to do basic scientific research. If you put the money out there for scientific research, scientists will compete on a peer review basis and get the grants. So I think there has been a realization on the Republican side. When you look even among Republicans, and the polling shows that 75-80% of Republicans believe in renewable resource allocation and funding, that some of our guys made mistakes last time, and I think we've been able to rally enough people to head it off at the pass.

Part of it, too, is a larger debate on environmental issues generally. When an appropriations bill gets passed, they're usually just a way to redistribute money, and then it's just the nickel goes here and a dime goes here and a dollar gets spent over there. You're not supposed to create any kind of legislation on appropriations bills. And we decided last time out we were going to take a whack at a lot of environmental issues we didn't like by them on spending bills. My point is, I think wetlands regulation in this country needs reform, but we should do it on a wetlands bill. We need to change the Clean Water Act and eliminate some of the foolish mandates that are in it, but do it in the Clean Water Act bill, don't do it in the appropriations process. So I think in a lot of ways we overreached on the environment last time out, and while the policy issues may have had merit, sticking them into a back room deal at the back of the deal was not the right way to do environmental policy in the United States. And if you've got to fix it, fix the underlying problem in the underlying bill.

Q: I am wondering what your views are on the success of these market acceleration programs like Green Lights, and what the relative support is in Congress for continued funding of those.

Klug: I'll be honest with you, I don't know that much about those programs. So, rather than have me tap dance up here and talk in circles I won't, except to say I think there is a continued federal role for energy conservation programs, and to cut those out would be a huge mistake. Again, I think it sends the wrong signals.

Q: We want to thank you for putting together the Renewable Energy Caucus with Congressman Schaffer, and I was wondering if there were any plans immediately or in the far future to incorporate energy efficiency activities in that caucus as well?

Klug: At this point we've really only had two purposes in mind. One of which was to do a renewables conference up on Capitol Hill, and included in that was a kind of a large, exposition display on renewables technology that was available, to have members of Congress kind of walk through and educate themselves so they had a sense of what programs are available. That was mission number one. Mission number two for Dan and I was to keep a close look on the energy and water appropriations bills, to essentially try to get this solved before we can get to a floor fight over continued funding for both conservation programs and renewables. And it looks like we're fine this time on the House side. We don't think we have any problems whatsoever. So we haven't really kind of gone to step number three, which is looking down the road. The first priority was obviously renewables, but kind of tucked in the background there has to be conservation. And I'm sure in the future that Doug and the rest of his colleagues from Johnson Controls will beat us over the head until conservation is intimately involved in our plans down the road.

Moderator: Well, this is going to be like last year, I can see. We are going to have some very divergent views on how to perfect our energy policy. Our next speaker is the Honorable Mary D. Nichols, the Assistant Administrator for Air and Radiation in the United States Environmental Protection Agency. I have had the pleasure to meet Ms. Nichols on several occasions, and I can say she has done a sterling job in her department in all the green programs, the Green Lights program, and the Energy Star program that everyone in this room is very familiar with. Please give a warm welcome to Ms. Nichols.

Mary Nichols: It's good to be here this morning, especially among these distinguished presenters. It's a real learning opportunity for me, to hear both these presentations and your questions.

I have three messages I want to offer today:

- First, to underscore the threats to public health in the environment from conventional air pollution and climate-changing greenhouse gases.
- Second, to review the key role of energy efficiency in reducing these threats, and our progress in partnership with many of you, carrying out the "Green Lights" and "Energy Star" programs under President Clinton's Climate Change Action Plan, and transforming markets for energy efficiency.
- Third, to emphasize the Clinton Administration's commitment to pursue pollution control and climate protection policies that will continue to build market conditions in which the energy efficiency industry will prosper, and to several specific initiatives that we at EPA are now undertaking with broad Administration support.

1. Air Pollution and Climate Change

To understand the importance of energy efficiency, one need only reflect on the many ways that energy production and use contribute to pollution. Much as we love and need electricity, one might say that the electrical generating industry is "multiple-pollutant challenged," being a major source of at least four of the most harmful pollutants:

- Nitric oxides, or NO_X, which contributes to ground-level ozone and fine particles, both causing serious health effects – thousands of illnesses and premature deaths each year. Ozone also causes crop and forest damage and NO_X contributes to regional haze, acidification of lakes and streams in the Adirondacks, and eutrophication of vital water bodies such as the Chesapeake Bay. At present, the 37-state Ozone Transport Assessment Group (OTAG) is seeking consensus on a plan to reduce NO_X from utilities and other sources as part of a region-wide program to meet the ozone standard across the eastern United States. Based on the best analysis to date, we believe summertime NO_X emissions from power generation in the OTAG region must be reduced by two-thirds, to below a million tons to achieve ozone standard. I'll tell you more about what the Clinton Administration is doing to achieve this objective a little later.

- Sulfur dioxide (SO_2) emissions play the largest role in acid rain. SO_2-derived sulfates make up a large fraction of health-damaging fine particles. A recent study attributed as many as 64,000 premature deaths a year to sulfate particles.

  Sulfates also contribute to the loss of visibility in national parks like the Grand Canyon, or Shenandoah.

  EPA's innovative, market-based program to control acid rain has been highly successful in reducing SO_2 even more quickly and cheaper than anticipated. In the coming years, we will have to consider making further reductions in SO_2 beyond those called for by the acid rain program, to protect against these health effects.

  EPA is reviewing the adequacy of both the ozone and the particulate matter standards and will propose any required changes in late November. We are working with a 60-member advisory committee of stakeholders from industry, environmental groups, state and local governments, and others on the formidable problems of making the transition to new standards while continuing to make uninterrupted progress on the necessary emission control measures.

- At high levels, Mercury is a serious neurotoxin. At lower levels, it has been linked to developmental impairment in young children. Mercury emissions settle in water bodies and bioaccumulate in fish. People are subject to significant exposure through fish consumption. Last year alone, 44 states had to issue fish consumption warnings because of mercury contamination. EPA has already included mercury control measures in recent standards for municipal incinerators, and we are working on studies required by the Clean Air Act before deciding on appropriate controls for mercury emissions from electric utilities.

- Carbon dioxide (CO_2) is the principal greenhouse gas contributing to global warming. The "Intergovernmental Panel on Climate Change" (IPCC), composed of more than 2500 of the world's best scientists, concluded last year that: "The balance of evidence suggests a discernible human influence on global climate," and that, "Climate change is likely to have wide-ranging and mostly adverse impacts on human health, with significant loss of life." Some of the likely impacts have almost Biblical proportions:
  - Higher death rates from more intense heat waves, and more severe air pollution.
  - The spread of diseases such as malaria, dengue fever, and cholera due to extended ranges of disease-carrying insects and other pests, and other ecological changes.
  - Regional threats to food production from drought and other environmental changes, sea level rise and increased storm damage.

At a minimum, climate change is likely to cost rich countries billions of dollars in damages and adaptation. In poorer countries, we could see famine and the creation of tens of millions of environmental refugees due to drought or sea level rise and storm damage.

President Clinton's Climate Change Action Plan is bringing about real reductions in emissions of CO_2 and other greenhouse gases that contribute to global warming. At the core of the Climate Change Action Plan, as you know, are EPA's and DOE's partnership-based "Green" and "Energy Star" programs. Focusing on the longer term, EPA is working with the Departments of State and Energy and other agencies in the international negotiations over targets, timetables, and policies that should be adopted for the years beyond 2000. These new obligations will be determined under negotiations to strengthen the Framework Convention on Climate Change between now and 1997. I'll return to this effort also in part three of my talk.

2. The Many Benefits of Energy Efficiency: Progress Report on the "Green" and Energy Star" Programs

It will come as no surprise to this audience when I say that energy efficiency – and EPA's energy efficiency programs in particular – pack a triple punch.

- Our programs and your products and services are saving money – billions of dollars – for the businesses, homeowners, schools, hospitals, and other that make use of them.
- They are making money, here in the U.S. and in the huge export market abroad – for the providers of these new goods and services.
- And they are preventing pollution, a central element of any long-term strategy to solve the challenges of air pollution and climate change that face America and the rest of the world.

Let me give you the latest figures. By the year 2000, the "Green" and "Energy Star" programs carried out by EPA and the Department of Energy under President Clinton's Climate Change Action Plan will:

- Reduce annual greenhouse gas emissions by 100 million tons of carbon dioxide – equivalent to eliminating the emissions of 15 million cars.
- Save over 100 billion kilowatt hours per year, resulting in savings of $6 billion to businesses and consumers. Public schools, and hospitals, and other government agencies will be saving over $2 billion in taxpayer money.
- Prevent over 200,000 tons per year of NO_X emissions.
- Prevent over 500,000 tons per year of SO_2 emissions. These reductions, achieved through profitable investments in energy efficiency, would otherwise cost U.S. business up to $50 million.

As many of you know, earlier this month EPA and DOE announced joint plans to expand the "Energy Star" program to cover many more products and appliances, and to use a common "Energy Star" label and promotional campaign. The "Energy

Star" label will be a direct line to consumers – a practical, continuous and reliable message to care about energy efficiency, and how to find it when you shop. We expect the "Energy Star" label to be on products and equipment worth hundreds of billions of dollars by the middle of the next decade.

These programs make sense in every way. As we make priority-based budget decisions at EPA, programs that bring the triple rewards of health, environmental, and economic benefits score very high, indeed. (Since I see that Congressman Klug will follow me to this podium, I can't resist saying that I hope these kinds of benefits are what he and his colleagues on the Hill are looking for as well.)

3. **New Initiatives to Control Conventional and Greenhouse Pollutants: The Link to Energy Efficiency**

Let me now return to the "supply side" – the power generating facilities that also generate the pollution we have been discussing. Many of you are concerned by trends in deregulation of the electric utility industry and the implications they may have for demand-side energy efficiency services.

The Clinton Administration is committed to achieving both the economic benefits of greater competition and the health and environmental benefits of cleaning the air and protecting the climate. If we pursue both objectives, I believe we also will strengthen market conditions for energy efficiency investments. So let me close by telling you of three areas in which the Clinton Administration and this EPA are moving forward with initiatives to limit emissions and internalize the health and environmental costs of power generation.

First, EPA is proceeding, with the strong support of the Clinton Administration, to achieve the large reductions in utility NO_X emissions (in addition to reductions in NO_X and other pollutants from other sources) needed to meet the ozone standards in the Northeast and Upper Midwest. We are working with the 37-state Ozone Transport Assessment Group (OTAG) in the effort to agree on a region-wide NO_X cap, with an emissions trading program modeled on the acid rain program to achieve the needed reductions as cheaply and flexibly as possible. OTAG is scheduled to make recommendations early next year, and to adopt implementing changes to State Implementation Plans in the following six months.

EPA has every confidence that the OTAG process can succeed. But if it does not, EPA is fully prepared to call for SIP revisions and issue Federal Implementation Plans as needed to meet our national health protection objectives. As expressed last week in statements by the President's Council on Environmental Quality, as well as the President's Council of Economic Advisers and the Department of Energy, EPA has the strong support of the Clinton Administration for this specific plan of action.

As some of you know, the statements of Administration support I just mentioned for our NO_X control panel were an outgrowth of concerns EPA had raised earlier this year over the potential emission consequences of the Federal Energy Regulatory Commission's "open access" rule. Under certain circumstances, EPA believes that there may be a substantial increase in low-cost power exports from older, heavier-emitting power plants, predominantly located in the Midwest. As late as April, when it promulgated the rule, the FERC continued to decline any role in preventing these pollution consequences. As a result, EPA referred the rule to Council on Environmental Quality under the Clean Air Act and the National Policy Act, out of concern for the potentially unsatisfactory health and environmental impacts the rules could have.

In a significant change of position, FERC agreed to reconsider emissions mitigation proposals if the OTAG process does not generate a timely and effective agreement.

What is the message of FERC's change of view and the Administration expressions of support for our NO_X initiatives? The message is that we are carrying out a unified program to meet both our public health and economic objectives. The OTAG and Clean Air Act process for meeting the ozone standards is not just an "EPA thing" – it's an "Administration thing."

Second, my office embarked on a broader, cooperative "Clean Air Power Initiative" – an effort to build a better, smarter framework for managing the multiple pollutants that come from the power generation industry. Specifically, the Clean Air Power Initiative provides the opportunity to develop a cost-effective reduction strategy for SO_2, NO_X, mercury, and (potentially) CO_2 because we believe these pollutants will be the focus of emission reduction efforts over the next 10-to-15 years.

Thus far, we have examined strategies for controlling the first three of these pollutants in the most cost-effective and flexible way – through a multi-state regional cap and trade approach.

Third, as I mentioned earlier, EPA is working with the State Department, DOE, the White House, and other agencies on developing new policies to combat climate change. As you know, the current climate treaty sets quantitative goals only through the year 2000. We are now beginning to develop the U.S. position on appropriate targets and timetables for the first two decades of the next century. Serious proposals advanced by other countries range from stabilizing developed countries' greenhouse gas emissions at 1990 levels in 2010 and beyond, to cutting their emissions by 20% as soon as 2005.

EPA and other agencies just held a major two-day public workshop to begin analyzing whether and how various targets could be met, and at what cost. Preliminary results of the government's analysis indicate that at least some of these targets could be achieved at modest and reasonable cost.

The key to reducing the threatened costs of climate change at a reasonable cost is to adopt policies that send clear and consistent signals to the market about the value of developing and adopting technologies that provide the products and services we want while reducing greenhouse gas emissions. Policies that emphasize continued application of energy efficiency technologies and continued improvements in technology and markets for that technology show strong net gains for the economy. Though the Administration is continuing to evaluate approaches and targets, and has not yet decided what policies to support in the climate negotiations, key decisions will need to be made early next year.

Success in these three initiatives will bolster market signals on the value of energy efficiency investments. We want and need your participation and your views in these three initiatives.

Those of you who are engaged in the business of making energy efficiency a reality, can feel proud to do what you do. You are doing well by doing good. I encourage you to market your products and services aggressively, for the benefit of the economy and the environment. Continue your development and introductions of new technologies – we'll do our best to send the right signals that help make a market out there. And let us know where you need our help. At EPA, we stand ready to continue to support your efforts to cut pollution and improve the economic bottom line.

Thanks very much for the opportunity to be here today. I'll be glad to take questions and hear your thoughts.

Q: What is the source of the mercury pollution from power plants?

Nichols: Well, all forms of combustion can cause mercury deposition, but in particular, different types of coals have different amounts of mercury in them. It is essentially the last, big remaining stationary source that has not yet had any controls placed on it under the Clean Air Act. But because of the way the Act is structured, the EPA is doing a major review of the whole issue of mercury and air deposition. We've got a draft report that's in front of our science advisory board now and being peer-reviewed, and we expect within the next year or two to have more definitive statements on that topic. But in the meantime, there is a good deal of information to suggest that mercury deposition from power plants will be an issue we will be looking at in the future.

Q: Could you address the issue of a gasoline tax please?

Nichols: Well, that's not within my area of jurisdiction, but I will say that there was a very serious effort which President Clinton launched as part of the Climate Change Action Plan to bring together the key sectors of the economy and the government that have dealt with the issue of growing vehicle miles traveled and the challenge of bringing the greenhouse gas emissions from personal transportation sector under some type of control. And we worked for over a year and did a tremendous amount of analysis on the topic, and achieved a great deal of consensus on what the drivers are and what the tools are for reducing those emissions. There was a general consensus that you had to deal with fuels, with the types of fuels and the quality of fuels, with the vehicles, and also with driving patterns. And without attacking all three of those simultaneously, we were unlikely to be able to achieve the goals of stabilizing, much less reducing, emissions from that sector.

However, there was no consensus between the oil industry and the auto industry on any of the measures that related to taxes or pricing of their products, or of further mandatory government action. And the basic situation, in my view, is that this would continue to be a stalemate unless and until there was a sufficient sense of urgency on the part of the industries that action would be taken. And frankly, threats on the part of the Administration, or suggestions of legislation, were not considered to be a sufficient threat to bring the major sectors of the industry together into any kind of consensus on the problem. I think the first step in the direction of achieving a greater sense that this is an issue that needs to be dealt with, really came from the IPCC report which has provided, I think, the strongest statement to date from the scientific community that this is an issue that is going to have to be addressed. I see in the industry sectors that I deal with, a growing recognition that this is a problem and that they will have to deal with it, but they want to deal with it, if at all possible, on an international and global basis, rather than through domestic initiatives alone. So we have a lot of work to do before we can come up with policies that will be successfully adopted, but it is not for lack of continuing to discuss the topic.

Moderator: Our clean-up hitter today is no one new to you. Christine Ervin is known by everyone in this group. She's the Assistant Secretary for Energy Efficiency and Renewable Energy at DOE. She has held this position since 1993. It's been, to put it mildly, a hot bed for her. Her budgets keep going down, down, down. And I'm sorry she's not here to lobby, but we will keep up the lobbying for her. In addition, Christine manages an array of programs, including new batteries, new automotive technologies, weatherization assistance, building technologies, state programs, and of course the very successful Federal Energy Management Program. Would you please welcome Ms. Christine Ervin.

Christine Ervin: Thank you, Doug, it's terrific to be here and I greatly enjoyed a number of comments from our previous speakers as well. Well I must tell you when I was offered the opportunity to give a spirited defense of the federal government's energy efficiency programs, I naturally hesitated. But after due deliberation, I decided to take you up on this. And I decided to take a provocative approach to laying out the very distinct points of view that we are facing this session.

I'd like to start by telling you a story that illustrates some of the current Congressional thinking about the subject of energy efficiency. Picture a time of great turmoil. A time like the French Revolution. A guillotine stands in the town square. The executioner drags a doctor down to the guillotine and straps him down. I told you this was dramatic. The executioner pulls the lever but the blade gets stuck and doesn't fall. The doctor shouts, "This is providence! I'm meant to live so I can continue to heal my patients! You must release me!" The executioner releases him.

Next comes a lawyer. The executioner straps him to the guillotine and pulls the lever but it gets stuck again. "This is a matter of precedent," the lawyer says. "You've let the doctor go and you must let me go." And the executioner releases him.

Next, the executioner drags in a certain member of Congress, straps him down, but the blade sticks again. The Congressman turns to the executioner and says, "You know, I could fix that."

Think of the guillotine as America's growing addiction to foreign oil, much of it coming from a very troubled Persian Gulf. Think of the guillotine as the danger energy efficiency poses to the global competitiveness of American industries. Think of it as the enormous potential for damage to the environment that Mary Nichols talked about, as nations like China and India undertake unprecedented economic development campaigns and decide on whether or not they will use clean energy or dirty energy. New energy technologies are the key to keeping that blade from falling. And in many cases, it is federal clean energy programs working in partnership with the private sector that are the catalyst to bring those good technologies into existence and into the marketplace.

I'd like to offer three observations about Congress' attempts to fix the guillotine, so to speak. Observation number one is that the assault now underway in Congress on federal clean energy programs really represents a sneak attack on the environment. The cuts being imposed on the programs will be just as devastating to the quality of the environment in our nation as attempts last year to roll back many of the key environmental laws and regulations. Observation number two: Those who argue that the federal government has no place in the marketplace, that the free market will take care of all of the nation's need to develop promising new technologies, are simply naive. They are operating in an economic fantasy world, and if allowed to shape national policy, their naiveté will make America a second-rate economic power in the next century. And observation number three: those who argue that these programs must be sacrificed in order to reduce the federal budget deficit, are saying in effect that we cannot afford clean air, we cannot afford good health, we cannot afford good new jobs, we cannot afford the kind of America that we want to leave to our kids.

Now to put these observations in better perspective, let me provide just a little bit of historical context, although many of you in this room, I think, are familiar with the roller coaster of the energy programs. As you know, the modern era of national policy on efficiency and renewable programs began after the oil crisis of the 1970s. Those crises dramatized the liabilities of our dependence on foreign oil, how our economy and lifestyles could be sent quickly into a tailspin at the whim of foreign oil-producing nations. Awakened by those crises, Congress and the President ordered the DOE and its predecessor agencies to undertake aggressive research into efficiency and renewable programs. All sectors of the economy were to become more efficient. And it's fair to say, as Congressman Klug did say, that some of that early money was misdirected. SynFuels may be a classic example. In some of the renewable energy programs, it was 100% federal money that went into the early, early technologies, often without a strong business plan with the private sector. But those were the very early years. And as a result of those early efforts, let us not forget that Americans have saved billions of dollars in energy and environmental costs over those past two decades.

However, during the 1980s, we know that the Reagan administration completely reversed course, and dramatically reduced the federal investment in those new technologies. And America lost its world leadership role in some technologies, like photovoltaics. A classic example. We invest, we start making headway on the technology, we pull back, and lo and behold, someone from overseas takes over the world's lead. It happened then. It's very predictable. Now the Bush administration, perhaps because it was forced to send our sons and daughters to fight another oil war, realized it was folly to let this nation drift into a swamp of foreign oil, so to speak. Working with a Democratic Congress, President Bush began rebuilding the federal government's programs in efficiency and renewables, a policy that has continued and has accelerated with the current administration.

The Congress elected in 1994 has begun the disinvestment process again. Earlier this year, Congress reduced the DOE's efficiency programs in government by a staggering 30%. Quite a blow to a budget that is relatively very modest. President Clinton has responded by proposing a Fiscal Year '97 budget which would restore funding to our '95 level. Not the '96 level which was higher, but about $1.1 billion in total. That's still well below the levels of the 1970s, too, I might add. It amounts to about $4 for every American, or the price of a fast food meal. Now, the House Interior Appropriations bill has proposed another 30% cut from the President's budget in these programs. And that's within the context of an overall House budget resolution that aims to eliminate the energy efficiency programs over the next two to three years. Not scale back, but to eliminate these programs.

Last year the public registered strong concerns about attempts to roll back environmental laws. And since then, as we know, the House Speaker has created a new environmental task force to provide a more balanced view, a more proactive view on environmental proposals. Well, if that's the case, I would submit that the task force has to take a very hard look at what sneaks through the Appropriations Committee. Now the Interior Committee's press release said at the very top that they were taking care of the environment. I'll say. Let me lift the fog from this issue. An attack on clean energy is an attack on the environment. Clean energy and clean environment are inextricably linked. Here's a little-known or talked-about fact. The production and use of energy causes more environmental damage than any other economic activity today. The problem is growing worse. The developing world is building new energy systems on an unprecedented scale, and unless clean energy technologies are available to them, their economic promise will mean major new pollution on a global scale. We can all but forget about stabilizing greenhouse gas emissions if we do not have a profound effect on the energy choices being made around the world today.

We don't have as tangible an energy crisis as we did in the 1970s. There are no gasoline lines to speak of. No cardigan sweaters in the White House. Today's energy problems are more subtle, but far more serious than those gas lines. Our energy use is changing the environment in destructive and long-lasting ways, like global climate change, the disappearance of rain forests and the species that occupy them, and the dumping of pollutants into air and water. Our job must be to find clean new energy resources.

Now here's another little-known fact. The DOE hosts the federal government's largest set of voluntary pollution prevention programs. And most of them are right here in the Office of Energy Efficiency. Make no mistake about it. When Congress votes against President Clinton's energy efficiency budget, it is a vote against the environment.

And I want to be real clear about the consequences of those cuts, too. When we talk about pollution prevention, it is not simply about making life safer for flora and fauna, although that's important. It's about public health, as Mary Nichols said. The American Lung Association estimates that we spend a good $50 billion/year in health care costs due to air pollution alone. Some 50 million Americans live in areas that regularly violate air quality standards, and as much as 80% of that urban air pollution is caused by transportation energy use. That's why we have a lead role in the federal government in developing clean alternative fuels, and developing cars that are three times more fuel efficient than today's. When we talk about preventing pollution with the lead role in the Office of Energy Efficiency, we're talking about preventing illness, cancer, and lung diseases.

Now, my second observation concerns the role of the federal government in the marketplace. We continue to hear the argument on the Hill, and among conservative think tanks, that the federal government has no business being involved in this so-called free energy market. One of the most vocal proponents of that view is the Cato Institute. Cato asks, "why is it that the DOE is so much smarter than any energy company, any energy executive, and any long-term investor?" I think the '97 House budget resolution clearly mirrors that unrealistic view of the marketplace. It says, "The budget resolution reflects the conviction that the nation's energy problems will be solved by the people in industries of this country, not by government spending."

I want to respond with a couple of reality statements. First, let's talk about the idea of a free market. I believe in the power and magic of the marketplace as much as anyone. I've taken a couple dozen economics courses, and I'm married to a professor of economics. Most of my best friends are economists, okay?! But any student of economics knows that in the real world, the marketplace is filled with subsidies and distortions and uncounted costs and benefits and information gaps and barriers. In the real world, cheapest is not always the best. Market forces of the moment do not always look after the nation's long-term security interests. For example, cheap oil may be great for the economy in the short-term, but insofar as it encourages us to use more imported oil, it is not good for the national economic security in the long-term, or for the balance of trade, or for the quality of our air. If the marketplace were infallible in its intelligent source of leadership, as Cato believes, then we wouldn't need a Congress or a White House or a Supreme Court. We wouldn't need governors or mayors or city councils. We wouldn't need conservative think tanks. The market would make all of the decisions for us; but in reality, it cannot. It is not a substitute for the intelligent leadership that protects and helps guide our long-term interests.

We just heard an economist joke previously. People who know me say that nearly every speech I give includes an economist joke. I wasn't going to today, but now I have to. And my favorite, it has been for months, is, how many economists does it take to change a light bulb? Answer one: None. If it needed changing, the market would take care of it. Or answer two: None. If the federal government would just leave it alone it will screw itself in.

The market purists make a couple of other mistakes, too. They think that we who operate federal energy research programs are trying to impose our efforts on the private sector, and they think that when it comes to developing new technologies, the Feeds can't do anything right. They're wrong on both accounts, and many of you know that. We do not attempt to replace the judgment of the private sector in America. Our programs empower the people in the industries of this country to develop and deploy the best technologies. For example, our Industries of the Future Program works closely with the nation's largest, most energy-intensive, most polluting industries in this country. Eight of them together account for 80% of the manufacturing waste in this country. By helping them prevent pollution, this research is going to help American companies save some of the $165 billion that they spend now to comply with environmental regulations. Industries drive that process. They create the vision for the technologies that they need in the future. They set the research agenda. We join them as partners. We facilitate it. The pulp and paper industry alone, for example, has estimated that our partnership with them is expected to save them 60% of the money they would otherwise save for complying with the Clean Air Cluster rule. That's a savings of about $8 billion for a program that is about $15 million of investment. Not bad.

I can give you an example in the building industry, too. No more critical industry perhaps in the country. In 1994, the construction/renovation business accounted for $850 billion in economic activity. That's about 14% of GNP. And it employs about 10 million Americans. But the building industry is tremendously fragmented. It is made up of more than one million construction firms, architects, engineers, developers, building equipment manufacturers, suppliers, building control manufacturers, window makers, insulation manufacturers, home builders . . . it goes on and on. Because of the fragmentation, it is very hard for the industry as a whole to collaborate on common research needs. In fact, the building industry spends less than 1% of all of its receipts on R&D, much, much lower than industry as a whole. And much lower than the building industry as a whole in European countries, for example. But when the DOE works in collaboration with the industry for a common agenda, you see some really impressive results. Building America, for example, has seen a new generation of housing that is 20-50% more energy efficient. It produces better indoor air quality. It recycles much more in the way of building materials. It sends much less construction debris to the local landfill dump. And, it costs less to build in the first place.

We have many new technologies that we've worked on. The electronic ballast is, of course, a classic example where a very small investment was made with the Lawrence Berkeley Laboratory and a couple of small businesses. And today, Americans have saved over a billion dollars in energy bills as a result.

Now, how about the claim that the Feds can't do anything right? The reality is that these technologies are winning a lot of awards. If you look at the nation's R&D 100 awards, which are almost the Nobel Prize for cutting edge technologies, the efficiency programs have won 6% of the awards with less than one-half of 1% of the budget. Thirty-one awards just in the last five years. Energy efficiency programs won last year's Discover Award for environmental technology, for producing the sulfur lamp, often called the third most important lighting technology since Thomas Edison created the light bulb. This year the Discover Award also went to a DOE program, for a device that purifies drinking water with ultraviolet light. A small, inexpensive device, created in large part at Lawrence Berkeley Laboratory, that's going to save the lives of many, many children around the world.

My third and final observation has to do with the federal budget deficit. Of course deficit reduction is an important national goal. President Clinton is committed to reducing the national deficit. But we've got to be selective and intelligent about how we reduce the deficit. It shouldn't be an excuse for cutting into our investment for the future.

Let me give you just a few examples of shortsightedness being done, so-called, in the name of cutting the budget deficit. It would cut our building technology programs by 40%. As a result, by the turn of the century, just four years from now, American consumers will pay $2.5 billion more for their energy. And by 2010, that's just 14 years, we will be paying $15 billion more each year on our energy bills for simply tens of millions of savings. The House bill cuts weatherization programs by 35%. Again, under this cut, 29,000 low-income families will be denied weatherization assistance in 1997. They will pay an average of $200 more in their energy bills. This is a population that already is spending three to four times as much on their utility bills as most of us do. The House will cut our Federal Energy Management Program by about two-thirds from the President's request. A budget request that's taken down from $32 million, a modest amount, down to $13 million. We will be spending hundreds of millions more on our federal energy bills to save tens of millions of dollars. I could go on and on but I won't.

But let me just say in closing: It is very difficult for me to understand what some of the Republican budget proposals really stand for. There are deep cuts to clean energy that, I think, clearly translate into more pollution and higher regulatory costs, into higher energy bills, fewer jobs, and the risk of spending much more in defense money to protect our national security. I think we might all agree clean energy used to be bipartisan. It needs to be bipartisan again, and our only hope, then, is for leaders from both parties to reject the theoretical extremes guiding the budget proposals and to come to reason.

We are beginning to see that reason being exercised in the House Renewables Caucus, under the leadership of Congressmen Schaffer and Klug, for example. And I'm encouraged that they are reaching out to the area of energy efficiency. It needs to be extended to the rest of the cutting-edge clean energy programs. For your business efforts and the national efforts. One of my favorite congressmen is Republican Vern Ehlers. Many of you may have heard him say that if BTUs were purple, energy would be the number one issue in America. Well, our job is to color energy purple, not to make it invisible. Thank you very much.

PAYING FOR GLOBAL ENERGY EFFICIENCY

Jefferson Seabright, Director, Office of Energy, Environment, and Technology, U.S. Agency for International Development.

Moderator: I met Jeff Seabright when he was working for Tim Wirth on Capitol Hill, at that time Senator Wirth. Before that, he worked for Senator Jay Rockefeller. Jeff did some pioneering work on environmental technology legislation while he was there with the senators. Prior to his Hill experience, he was a foreign service officer for the Department of State and served in Brussels on the international staff of the North Atlantic Assembly. Prior to that, he worked with the Congressional Research Service, and he has an MS from the London School of Economics and his undergraduate degree from Oberlin.

Jeff Seabright: Thank you, Jim, and I would like to thank USEA and Johnson Controls for inviting me to share some thoughts with you today. I see a lot of friends here in the room and a lot of partners in our work. Before starting, I would like to specifically thank Jim Sullivan, who really has provided over many years at USAID the leadership that brought this Office of Energy to the forefront of the international energy community. During Jim's leadership USAID was able to bring its energy efficiency support and work to over 35 countries around the world, constituting the largest bilateral donor program in energy of any of the donors. I would like to thank Jim for that work.

Unfortunately, our ability to provide that leadership, going into the latter part of the 90s, is certainly under attack. Our mission and our budget have come under assault, and I don't know if any of you saw the *Washington Post* article yesterday, but the US now has the distinct honor of being dead last among the developed countries in terms of the share of our GNP that we commit to sustainable development. The average Portuguese family spends more on development assistance abroad than the average American family. The average Irish family spends more. We also have dropped to fourth in terms of total effort. Germany, France, and Japan now outrank us in terms of international development assistance. I think this is unfortunate, because it not only means that we are not able to provide the leadership that the world needs as we head into the next century, but it also means that the influence that comes with American development assistance, and the market-opening measures that that brings for US business, will also decline. So with that little advertisement and note, I will turn to the focus of this panel.

I would like to do this in two parts. First to make some general comments about where we are now at USAID on energy efficiency, and then focus specifically on the energy efficiency finance analysis that we are currently undertaking, and which should be available some time in the next month. USAID currently has major energy efficiency programs in Asia, Latin America, and in the Newly Independent States of Europe. More specifically, we have major activities in Brazil, Mexico, Egypt, Morocco, India, Philippines, Russia, Ukraine and several Eastern European countries. Our focus is really on the policy, technical assistance, and training aspects of supporting energy efficiency work in these countries, not on the provision of project funding. We are not OPEC, we are not Export-Import Bank, we are no TDA. We don't have the resources even if we wanted to play that role of investing in specific projects. Rather, we try to make an upstream investment in getting the markets right and in working through our technical assistance and training

Clearly, around the world what we are seeing is a transformation in terms of what is going on with privatization in energy sectors. Many market barriers are coming down, institutional impediments that previously posed insurmountable obstacles are beginning to disappear. And I think this poses significant opportunities for energy efficiency investments in developing countries. Our analysis shows these markets should be growing substantially in the future, and that finance does represent a problem, which is the subject of the panel discussion this morning.

USAID has, through its energy efficiency project, begun to focus on the finance issues. We are targeting not just the policy issues, one aspect of which is finance, but we also are working to leverage the investments that others are going to make. For example, in Brazil, the World Bank is making a very large loan, I think the largest loan package the World Bank will do as a free-standing loan for energy efficiency. For another example, Mexico has an $80 million DSM program, and we are able to use some of our resources to work with them on their national program, open their market, and bring benefits back to US businesses interested in that market. We are also working to support the promotion and creation of Energy Service Companies (ESCOs) in developing countries, as a means of bringing some energy efficiency entrepreneurship to the marketplace. And we are also working on private fund development, not actually funding these efforts, but working to catalyze private sector funds on energy efficiency. In Brazil, for example, we are working to bring together a $20 million energy efficiency private sector fund for investment in Brazil.

Turning now to this analysis we are engaged in, there are several areas that our report will be focusing on that address opportunities for improving finance for energy efficiency investments. Aggregation of the market is a key element. But one of the things we believe is that the market is really very fragmented, and this is no surprise to any of you. The number of stakeholders, the different end-users, the different technologies, mean that this is a highly fragmented market, and that makes it difficult to aggregate efficient investment in this sector. There are problems with public/private partnerships in getting the kinds of working teams put together that are necessary to bring some of these things off, and fairly minimal participation on the part of commercial lenders in many developing countries, and ongoing problems with policies that are key to the whole enterprise.

So I think the question for this panel, where will the money come from, does not have one answer. The answers will come in various shapes and forms, depending upon the circumstances and the opportunities. We at USAID we are working on six areas of improvement for increasing the availability of financing for energy efficiency in developing countries.

The first, and perhaps the most obvious, is increasing the participation of commercial credit. Obviously in the US, this is the single largest source of financing. It is not the case in developing countries. Leasing and term loans represent powerful tools for energy efficiency financing. Our information is that in Korea, for example, the potential just for lease financing for energy efficiency, is something on the order of $6 billion. In Indonesia, over $3 billion; in Chile over $750 million. So it is an enormous market and obviously has a lot of advantages, such as minimal initial cash requirements, and the possibility of structuring positive cash flow. Commercial lenders, in addition to providing direct credit, can also play other important roles in developing countries: as guarantors, fund administrators, and intermediaries.

The second area has to do with improving funding sources for Energy Service Companies (ESCOs). Again, Energy Service Companies are very prominent in the US, but outside the US, the opportunities for energy efficiency entrepreneurship are somewhat limited. One of the reasons for this is the lack of access to financing. We hope that by working to strengthen the role of ESCOs in developing countries through technical assistance, training, and other means, we can help increase the availability of finance and activity and energy efficiency through support for Energy Service Companies.

The third area that we have identified is a focus on vendor financing. Obviously this is an extremely useful application for large markets, for commonly used equipment where there are large numbers of end-users. Efforts to increase the amount of vendor financing available for energy efficiency in developing countries could significantly impact the markets.

A fourth area, not surprisingly, is the role of utilities in energy efficiency financing, although clearly in most developing countries, the role of utilities is undergoing a dramatic transformation as these energy economics privatize and become unbundled. That poses some significant challenges. However, I think in a competitive market you can find many interested companies abroad, just as they are here. I've had discussions in the Philippines, they are very keenly interested in DSM as a means to retain customers. They are looking at a potentially competitive environment and they don't want to lose large industrial consumers, and they are anxious to find ways to help them save energy and reduce costs. So there are many different roles utilities can play in financing energy efficiency, from facilitation to collection to being a financial services provider, and I think that this is an area that requires some ongoing work.

A fifth area is the establishment of country- or region-specific funds targeting the specific end-use which allows project evaluation to be standardized. To target end-users of similar creditworthiness, to reduce transaction costs, and to create sufficient capital demand so that the financing is justified. And to assist in bringing together a finance program that can be made available effectively to members, to advertise that and to help organize the market through the creation of a fund similar to the KEMCO Fund in Korea.

The last area that we've focused on is looking forward into the future in terms of developing new and innovative mechanisms for promoting energy efficiency. We've identified at least three areas. One is emissions credits and a global carbon market. This is not an area that currently seems to have a lot of interest by the marketplace, but joint implementation credits potentially could become an important source of financing for energy efficiency investments if that market ever were to materialize. A second area that is quite interesting is to link energy efficiency financing to independent power producer financing. Making bids on generation and including energy savings as part of an overall bid, is an interesting potential for future financing and market development. And lastly, taking a look at tapping secondary markets for accessing new sources of financing for energy efficiency, something that people at DOE have been looking at and working on closely here in the US. It is probably premature to look at secondary markets in developing countries until you have a primary market, but it is certainly an area, in terms of standardized or aggregated markets, where you really do have something to trade, and where a secondary market could help free up scarce capital for energy efficiency investments.

So that is a broad outline of the report that we have been working on over the last several months at USAID on the specific subject of overcoming obstacles to financing energy efficiency.

ENERGY EFFICIENCY PROGRAMS IN A DEREGULATED INDUSTRY

Michael Baly III, President & CEO, American Gas Association.

Don't look to the American Gas Association for the definitive word on energy efficiency programs under deregulation. Not yet, anyway. While unbundling (or restructuring) of pipelines is complete, the full impacts of competition on the structure and content of gas retail services have not been determined. The decoupling (or unbundling) of gas sales from distribution and other utility services is new for the vast majority of utility customers, the small commercial and residential consumers.

As utilities work with regulators on unbundling plans, however, they are evaluating currently mandated energy efficiency programs, and some trends are emerging. Early regulatory actions signal flexibility in recognition of competitive changes. Some state commissions, in fact, already have eased or eliminated mandates for such programs.

Some industry observers have concluded that these actions portend the end of energy efficiency programs at utilities. Far from being a casualty of deregulation, however, programs that help customers manage their energy bills and get the most out of energy technology will be important to sales in gas markets.

Will all energy efficiency programs now mandated by utility regulators survive the competitive transformation? Unlikely. Which programs survive, how they are marketed and by whom will depend on a number of factors. But competition will lead, in my opinion, to greater innovation in the sale of energy efficient products and services. Those preparing themselves to compete in energy markets recognize the sizable opportunities for sales of energy efficiency-related product and service offerings.

My optimism about the future of efficiency programs in a competitive environment might appear paradoxical when one looks at their history under regulation. While the American Gas Association has always promoted efficient fuel use, numerous mandated energy efficiency programs have been the subject of controversy, especially as they have been structured in recent years under the regulatory umbrella of demand-side management (DSM).

DSM was originally developed for electric utilities – and was applied later to gas utilities – as a method of getting efficiency-improving products into the marketplace and the hands of consumers, who otherwise might not have bought them. The rationale was twofold: 1) that consumer costs would be minimized through such programs, because reducing a utility's need to add capacity that would generate electricity at higher cost than existing generation sources will, in the long run, save money, and 2) that energy conservation is good public policy, Standard tools of DSM were rebates and other financial incentives for energy efficient equipment and appliances, free energy audits and weatherization materials.

The DSM concept generally was accepted by electric utilities because the price tag on DSM programs was far more attractive than those associated with constructing new power plants. Not so for the gas industry. Unlike many electric utilities, gas utilities can generally meet increased demand on their systems without expensive construction; whether consumption stays the same or is increased, whether new homes and buildings are added, gas lines are generally in place to serve them.

The National Regulatory Research Institute (NRRI), after evaluating mandated DSM programs across the country, pointed out this basic deficiency of gas DSM when it concluded that the marginal cost of increased gas consumption, "...includes little more than paying for the gas itself, its transmission and, in some cases, putting another pipe in the ground."[1] Another economic evaluation of gas DSM by a team of experts associated with the Gas Research Institute compared the benefits of gas DSM programs with program costs. Their conclusion: "Avoided cost estimates from across the United States prove that most, if not all, non-weatherization (gas) DSM is economically infeasible."[2]

---

[1] The National Regulatory Research Institute, Unbundling the Retail Gas Market: Current Activities and Guidance for Service Residential and Small Customers, (May 1996), p. 80.

[2] Xinying Wang, Larry Dombrowski and Daniel Hamblin, Will EPAct Kill Gas IRP?, Public Utilities Fortnightly, (April 15, 1995), p.45.

Gas DSM programs also have been criticized for distorting market signals through their most common method of financing – subsidization by other customers. In the regulated environment, a utility generally recovers DSM costs in its rates. Customers who benefit from program incentives cut their consumption through efficiency measures and experience an overall drop in their gas bills. All other customers on the system, however, find their costs go up, with no increase in consumption. Another market distortion results from DSM programs when decreasing consumption leads to the under-utilization of a system, creating greater costs per unit for consumers than full utilization.

For these reasons and others, in 1993 the Illinois Public Utility Commission declared DSM (and gas integrated resource planning, in general) "a clear loser," and refused to mandate DSM programs for gas utilities. Citing similar arguments, commissions in Colorado, North Carolina, Arkansas, West Virginia, South Dakota, and Tennessee refused to order DSM for gas utilities.

The controversy over DSM continues as the industry moves toward greater competition, and with good reason. In regulated markets, customers have no choice but to pay for mandated conservation programs, whose costs are included in their rates. In a competitive environment, where DSM programs raise the costs of gas service, some utility customers will choose a non-utility provider who is not under similar mandates. When customers leave the utility system, the – now fewer – remaining utility customers assume the entire DSM cost burden, and the utility's competitiveness is further weakened. Clearly, regulatory changes are required if utilities are to compete in deregulated energy markets.

Recognizing this, the NRRI, earlier this year, noted, "Selling below cost or subsidizing certain products or services is rarely seen in non-regulated markets...An LDC operating in a competitive marketplace will behave more like the firms just described. Its ability to offer subsidized services, although not completely erased, becomes greatly diminished."[3] (emphasis added) The report goes on to predict regulatory response to competitive changes, saying, "In sum, the accelerated competition induced by service unbundling would make it more difficult for commissions or legislatures to require LDCs to undertake non-market social activities."[4]

---

[3] The National Regulatory Research Institute, op. cit., p. 79.

[4] The National Regulatory Research Institute, op. cit., p. 81.

Many utility regulators, reflecting this thinking, are waiting to see how utilities restructure and how competition evolves before clarifying rules for energy efficiency programs. Some are allowing utilities to put conservation programs on hold while they focus on restructuring their companies. Massachusetts recently granted utilities permission to exit mandatory conservation programs, identifying DSM as an impediment to competition in a deregulated environment.

These actions do not mean that energy efficiency programs are not valued by the American Gas Association or by regulators. Energy efficiency will be pursued by our membership precisely because of its significant value. But some changes will be made concerning how programs are offered, by whom they are offered, and how they are financed and by whom. A key determinant of program restructuring will be how a programs stands up to market and social tests.

<u>Market Test:</u> Energy efficiency programs that meet the market test will be 1) attractive enough for the customer to buy, 2) cost effective enough for an energy service marketer to promote, and 3) efficacious in achieving energy conservation. Those programs that satisfy these criterion will be offered by energy service providers, regulated and unregulated alike.

Among them will be energy management consultants, who might limit their sales to conservation-related information. Other marketers will sell many, diverse energy-related services. The focus of early marketing activities will be the commercial and industrial markets, as well as large aggregates of residential customers, such as subdivisions. These customers have the greatest potential for energy savings, and thus, cost savings.

Many gas utilities will offer unmandated energy efficiency products and services to the extent that sales can be profitable or, at least will not increase rates to consumers. (Helping consumers cut their energy bills has significant market value that could represent the difference between retaining and losing customers.) The level of utility involvement in energy efficiency programs will be determined by a number of factors, including the strength of competition faced in a particular service territory, the need for additional peaking sources or the existence of excess capacity, consumer demand, and the utility's ability to finance or employ innovative financing mechanisms to enhance conservation-related sales.

Corporate goals, or the utility's vision of its market position, also will determine a utility's conservation offerings, and some utilities will decline to offer any unmandated sales and service functions that are not related to the distribution of gas. These utilities or their holding companies are making decisions now (if they haven't already) concerning the creation of subsidiaries to perform these non-distribution functions. Some recently formed subsidiaries are implementing innovative approaches to participation in the conservation or energy management services market, offering products and services as diverse as weatherization, architectural design, remodeling and retrofitting services, high efficiency equipment and alternative energy products, such as solar panels.

Non-utility subsidiaries will have the option to sell fuels in addition to natural gas, such as electricity, propane and/or oil. Fuel-neutral energy service providers and their customers will evaluate energy sources for their efficient applications, availability and market price vis a vis other fuels. Fuel flexibility allows energy companies to meet customer needs with the right energy mix at the best current prices.

This development is significant in its potential to correct economic distortions resulting from the narrow DSM focus of single-fuel use. Under DSM, for example, a capacity-squeezed utility could offer substantial incentives to reduce consumption and thus ease its capacity strain. At the same time, however, such incentives could detract business from an underutilized, competing utility in the area. Single-fuel conservation planning also is blind to overall energy policy goals such as reducing the country's trade imbalance, enhancing national security and cleaner air. The competitive market and the emergence of fuel-neutral energy service providers could contribute greatly to achieving these larger goals.

Social Test: Some conservation programs may fail the market test, but pass social policy tests for their potential contributions to achieving conservation objectives. When this occurs, we can expect government to intervene to ensure these objectives are met. We can hope their interventions will circumvent the negative consequences of energy efficiency mandates, namely negative rate impacts for consumers and competitive disadvantages for utilities.

The National Association of Regulatory Utility Commissioners (NARUC), in its 1995 study of gas integrated resource planning and DSM, pointed to several options the government has for meeting conservation objectives without imposing mandates on utilities. Among them are appliance efficiency ratings and building code standards, which can be effective in cutting energy consumption.[5]

One of the other ways to encourage efficiency is to increase the amount of consumer information available on the energy consumption of various appliances. Under the federal government's appliance labeling program, for example, consumers are now able to compare the annual energy consumption and cost of appliances powered by different fuel sources – this practice ensures that those consumers can make an "efficient" choice.

From a public policy perspective, the American Gas Association has long supported programs that take this practice a step further. When it comes to funding research, promoting new technologies or building new government installations, we believe that policy-makers need to consider the source-efficiency of a product, not just its site-efficiency. In other words, they should examine the total amount of energy consumed during production, transmission and end-use. This type of analytical approach would also serve consumers well – as long as the information was readily available to them.

In conclusion, energy efficiency programs will remain a part of the American landscape because conservation is in the interests of consumers and energy providers alike. Consumers want efficient homes, offices and production facilities, and marketers preparing to compete in new energy markets know that promoting energy efficiency is essential in a customer-driven market.

Competitive energy markets could produce, perhaps, the greatest innovations ever achieved in energy efficient products and services. In order for this to happen, regulators and federal agencies must be careful not to competitively disadvantage any marketers, including utilities.

---

[5] National Association of Regulatory Utility Commissioners, Gas Integrated Resource Planning and Demand-Side Management: A Compendium of Case Studies, (November 1995), p. iv.

Regulators and federal agencies dealing with energy issues might do well to call a time-out from mandates, thus providing themselves with the opportunity for a fresh look at existing and proposed conservation programs. Programs must be evaluated for their contributions to national and local energy goals. Existing programs must be scrutinized for the results they've produced and the potential costs and savings associated with the programs. If a conservation program raises consumer costs without any significant benefit to the consumer, the program should be eliminated.

A time-out also would give the energy industries time to make the adjustment from the largely regulated to the deregulated market environment. There is much reason to believe that given some latitude, the market will be innovative and effective in selling conservation, but it must be given time to do so.

Energy efficiency programs that do not distort market signals and do not hurt the competitiveness of the provider will be very attractive to gas utilities and, increasingly, non-utility services. I echo the views of many of my colleagues in the energy industries when I say that energy efficiency becomes a customer service in a competitive environment. Energy service providers that can differentiate themselves for being responsive to customers, for innovative service and best cost, will be tomorrow's success stories.

ENERGY EFFICIENCY PROGRAMS IN A DEREGULATED INDUSTRY

Robert Culshaw, Minister-Counsellor, Trade and Transport, British Embassy.

Britain embarked on a program of privatization in the 1980s which is still in progress today. There was from the beginning a strong practical thrust behind this program. There are two key elements. First, governments are not, and never have been, good at running businesses on a day to day basis. Second, economic efficiency is best served by competitive markets; consumers and producers alike reap benefits from competition, which is the best way of balancing supply and demand in the long term . In the light of their experience with earlier privatization's the British Government came to believe that this approach should apply to energy just as much as to other commodities.

The electricity industry before privatization, like many other utility industries, had a monolithic structure. The Central Electricity Generating Board (CEGB) combined transmission and generation functions in one huge entity under the ultimate control of Government, along with several regionally based supply and distribution boards. This meant that Government could intervene in economic decisions about what type of plant should be built, when and where they should be built, and the final price at which electricity was sold. These decisions could be (and were) influenced by political considerations, with little regard to the consumer and taxpayer who had to foot the bill.

After a preliminary attempt to introduce some competition in generation in 1983, the British Government decided in the late 1980s to put an end to the possibility of its own intervention, and to force the energy sector to evolve and innovate by exposing it to the kind of disciplines which normal companies face in the commercial marketplace. This is how restructuring and privatization of the electricity industry began.

The Electricity Act which was passed by the British Parliament in 1989 allowed for the restructuring of the CEGB into separate transmission and generation functions. Generation was to operate in a competitive market, while the transmission element, considered to be a natural monopoly, was to be subject to price regulation. An independent regulator was appointed specifically to look after the electricity market and the needs of electricity consumers. The area boards which were responsible for supplying and distributing electricity to the final consumer were replaced by 12 private Regional Electricity Companies known as RECs, who were to face a gradual phasing in of competition on their supply business and continuing price regulation of their monopoly distribution network.

The process of liberalization began in 1990 and is not yet complete. The modern nuclear generating stations were only privatized in 1996 and the final stage of the privatization process – the extension of choice of electricity supplier to some 22 million domestic customers – will take place in 1998. So the British electricity industry is still in transition. But already there is evidence of falling prices, higher standards of service to customers, and significant environmental improvements: the last of these may be particularly interesting in the United States. The liberalization process has brought about a massive change in attitudes as the market has delivered strong signals about the best way to allocate resources and as competition has stimulated improvements in efficiency.

The Generation Market: If we look in more detail at one part of the industry, the generation of electricity, we can see how this change in attitude has brought about impressive achievements. In the new system contracts with suppliers are made and plant is called to run by the National Grid on the basis of availability and price, all in a transparent way. Generators face real pressure to fuel and run their stations efficiently, to identify where the real costs lie, and to work out which costs can be avoided and which minimized. Each station must maximize its contribution to the company on the basis of availability, thermal efficiency, staff levels and costs: otherwise it will face closure.

Since privatization National Power has been the largest of the generators. Comparison of its performance in 1990/1 with its performance in 1995/6 shows that productivity increased by a staggering 140%, operating costs were down by 23.5%, and profits were up to $1200 million. This success was achieved despite a 10% drop in turnover as market share was lost to new entrants. What lies behind these statistics, and how have these efficiency savings been made?

The changes at National Power's flagship power station, Drax, show how competition has focused attention on costs and value for money, and so has helped to produce these impressive figures. Drax is a 4000 MW coal fired station which was completed in 1986. It is the largest station operating in England and Wales, able to produce enough electricity to meet the needs of around 4 million people, i.e. a metropolitan area the size of Boston. After privatization in 1991/2 its controllable costs were assessed at $110 million. In 1995/6 the implementation of a series of measures had reduced this to $70 million. Staff reductions accounted for a substantial proportion of this saving. What took in excess of 1200 staff to do can now, through better working practices, be done by half that number. Delayering, team working and the removal of restrictive practices have all played their part. Despite staff reductions the plant is running better than ever, with plant availability up and an increase in output. Better marketing of the station's by-products has also contributed to increased profits.

Such savings, and increases in productivity, can be seen right across the generation market. National Power's halving of overall station operating costs is by no means exceptional. PowerGen, National Power's main rival, has since privatization reduced the numbers required to operate its 2000 MW stations from 750 to 250. Staff in its core business are down even more, from 10,000 to 2,800 in 1995/6. Productivity in the UK business has reached a new high of 18.4 Gwh per employee – an increase of 110% since March 1991. Much of this is due to extensive benchmarking against leaders in the field. PowerGen launched a benchmarking project in 1992 to bring its five 2000 MW coal fired stations up to the level of world best practice. A number of countries including the United States were examined against a checklist of key performance indicators. Finally a small team undertook two exhaustive rounds of detailed comparisons against the most competitive stations in the US, visiting seven plants from Ohio to Texas. The goal was to blend good technical performance with low operating costs, while maintaining standards of safety and environmental protection. PowerGen has learned important lessons from its benchmarking exercises, not least the importance of flexibility, of simplified staff structures, and of new approaches to fuel purchasing, maintenance and contracts. By 1997/8 PowerGen expects its stations to meet the world's best practice.

Similar benchmarking exercises are taking place in all areas of the privatized utilities. Privatization has unleashed the industry's innovation and increased its determination to achieve ever higher performance, to the benefit of its shareholders and its customers. The companies now focus on costs in order to import productivity, competitiveness and profitability. As the competitive pressures intensify and opportunities for expansion and diversification in the home market diminish, these British privatized companies are increasingly selling their expertise abroad and investing in projects in a wide range of countries.

Investment: National Power has already invested over $1200 million overseas, including Portugal, Pakistan and Australia. In the United States, it has as acquired the American generating company, TEVCO, now renamed American National Power, and in summer 1996 the 158 MW Milford power station in Massachusetts. Other recent acquisitions include the Hazlewood Power Corporation, which runs a 1600 MW power station and coal mine in Australia, and a 26% stake in the Kot Addu plant in Pakistan. Similarly PowerGen led a consortium which acquired the Australian power company, Yallourn Energy Ltd, for $1.8 billion: this company is the first of five newly created generation companies to be sold in the state of Victoria. PowerGen now has committed investments in new business activities of some $1.7 billion. It has invested in a 1220 MW coal-fired plant at Paiton, Indonesia which it will operate and maintain. It is also pursuing other opportunities in Indonesia as well as India, Thailand, the Philippines, Hungary, Latin America and the Middle East. Last year it won a contract to carry out a study for the major overhaul of a coal fired power complex in Kazakhstan, where it aims to double the station's capacity.

This kind of investment in the electricity industry does not flow all one way. While British companies are investing overseas their US counterparts are seizing opportunities in the UK. There have been no less than 3 mergers of Regional Electricity

Companies with US utility companies in the 1990s, as well as substantial investment with UK partners in generation joint ventures in Britain. In addition Edison Mission Energy, a company owned by Pacific Light and Power, has purchased the pumped storage business which previously belonged to the National Grid. Such cross utility mergers and diversifications have led to some development of the industry structure put in place at privatization. As the scope for savings from economies of scale increases with competition in the domestic supply market in 1998, further new utility combinations may develop. Already the renaming of the companies shows the way ahead. "Eastern Electricity," for example, are now known as "Eastern": "Yorkshire Electricity" are "Yorkshire Group." This reflects the fact than many RECs have diversified into telecommunications and gas marketing. Some, such as Norweb, SWALEC and Scottish Power, have joined forces with water utilities. This trend is likely to continue.

The Supply Market: In the generation market competition was introduced from the start, and has had time to bite. In the supply of electricity to final consumers, competition has had to be phased in gradually because of the vast numbers of consumers in the market. Initially only those taking over 1 MW could choose their supplier. In 1994 this choice was extended to the over 100 kw market, and in 1998 the remaining 22 million domestic customers will be able to choose their electricity suppliers. The franchise reduction in 1994 led to significant price reductions and work is now underway on the last and biggest franchise reduction.

What will the benefits for domestic customers be? Some commentators suggest that, because the margins in the supply business are not as great as for generation, the price reductions for consumers may not be large. But as suppliers examine their costs to try to increase margins, they will put pressure on their suppliers (the generators) to trim their own costs even further.

As this quiet revolution in the supply market gathers pace, the RECs themselves are cutting costs and restructuring to meet the challenges of 1998. East Midlands Electricity, a Regional Electricity Company which serves 2.2 million business and domestic customers in the center of England, is now closely focused on its core business. An initial phase of sporadic and unprofitable diversification was followed by a change of strategic direction in 1994. A new management team divested the company of every business not closely related to its core competence of supplying and distributing electricity. In a move typical of the industry East Midlands Electricity restructured its business and reduced its 13 district offices to 6 regionally based business units. Staff numbers reduced steadily as responsibility was devolved downwards.

Yet East Midlands Electricity, and RECs like it, are not cutting costs at the expense of the consumer or of customer service. Two linked customer service centers are being established in East Midland Electricity's area and the average response time for incoming calls is 5 seconds (for comparison an emergency call to the British police rings for an average of 18 seconds). Other improvements include fixed time appointments booked immediately on screen by customer service staff, and bespoke software to help staff choose the most appropriate engineer according to experience and location. These are all signs of the new emphasis on low cost solutions and high customer service which has resulted from competition and effective regulation. Standards of service are closely monitored by the regulator and he has issued increasingly tough performance standards for all the RECs. These performance measures, and the performance of the RECs against them, are published annually. Almost all the RECs are investing heavily in their customer service centers. They know that a happy customer is cheaper than an unhappy customer.

Achievements: Companies have increased profits and improved service to their customers in generation, transmission, distribution and supply. But the key measure of success of any privatization is its effect on prices. These have fallen dramatically.

It would be foolish to pretend that everything has always gone according to plan. Any privatization is a huge undertaking and it is impossible to foresee the consequences of every action taken. But the regulatory regime has been effective: Britain now has a legislative framework which can cope with the unexpected in a transparent way, and so maintain market confidence and stability.

The United Kingdom is rightly proud of its achievements. But there is no complacency. Britain's privatized electricity regime has formed the basis of many privatization's around the world, so the competitive instinct has been released. British electricity companies know they must continue to improve, both at home and overseas. They are confident that they can.

ENERGY EFFICIENCY PROGRAMS IN A DEREGULATED INDUSTRY

Frederick Buckman, President and CEO, PacifiCorp.

Competition is transforming the electric utility industry, from the way we market power to our customers, to the manner in which we operate our plants, to the management of our energy efficiency programs.

Yet, a common theme underlies every change: utilities will have to do everything smarter, or they won't survive. A major shakeout will hit our industry, and you'll find the utilities left standing will be those that exceed their customers' expectations and develop solutions in innovative, cost-effective ways.

Finding new ways to fund energy efficiency is one of those challenging issues, and PacifiCorp is committed to this effort. Yet, to understand some of the approaches we are exploring, it's necessary to touch briefing on the competitive challenges we are facing.

The wholesale energy market has become, at least in the western United States, a fully competitive market. Virtually no new sale is being made on a cost-of-service basis. And the market is not only competitive, it is very liquid.

Many of our customers are finding ways to create choice for themselves, either by generating their own power or forming municipalities that can buy their own bulk power. And customers are using this power in negotiating with us for rates and services.

Customers also are clamoring for the right to choose their energy provider, and it's very likely that we'll see more and more states pass legislation that lets them do this in the next few years.

At PacifiCorp, we're in a good position to compete. As the third largest utility in the west, we serve 1.3 million customers in seven western states, and we provide telecommunications services in 1 states. We recently expanded our reach outside the United States and began providing electric service to 540,000 customers in

Australia. We are one of the nation's lowest-cost producers of power, with average electric rates of 4.5 cents per kilowatt hour.

But even with a solid track record and clear vision, we face significant challenges. We need to be able to offer the lowest priced, most dependable power and value-added energy products and services. And every utility will be competing with us to do the same.

Given this new competitive environment, some people fear that energy efficiency programs will suffer, because often they are subsidized. I don't think that's the case, but conservation must be done differently.

That's because our industry has suffered from a disease that I will characterize as "you really otta wanna." Utilities have been told for years: "You really otta wanna sign up these independent power contracts even though they are .06 kilowatt/hour contracts in a .03 market. You really otta wanna install conservation programs on customer premises, whether they make sense or not. You really otta wanna subsidize one set of customers with another set of customers."

In the past, we could live with that disease, because the "you really otta wanna" was a discussion among the utility, the regulatory commission and interest groups. Generally, public utility commissions were supportive of conservation and allowed utilities to roll the costs into rates. The people who paid the price for this disease were the customers who were not at the table.

But now the customer is saying, "You really otta wanna give me a choice. You really otta wanna compete, because I want competitive behavior to be the kind of behavior from which I buy. And you really otta wanna be our business partner rather than just someone who sells us a commodity."

I believe that the customer is right. Moving in a more competitive direction is good public policy. PacifiCorp is advocating that all customers be able to choose their energy supplier by the year 2001. We also think customer choice should be unfettered by issues of transmission access, stranded investment or local problems. I think we should look to local solutions rather than national ones, although it may take federal legislation to achieve some consistency.

In the new competitive marketplace, I believe that you will see more demand rather than less demand for efficiency. You will see new energy facilities able to produce energy at a total cost that is less than marginal cost of existing facilities. They will do this by being dramatically more energy efficient.

You will also see innovative solutions to energy transport over long distances. In our system, for example, we have found that if BC Hydro wants to sell energy to Arizona Public Service, we can sell 1,500 miles of transmission line to make that transaction. But the fact is, we can make the transaction by ramping down energy production at a plant in Arizona and no electrons will flow from Washington to Arizona. So there is no line loss to deal with.

Competition also will spur utilities to increase their energy service offerings, and that will include providing low-cost solutions to customer energy needs. This energy efficiency will be driven by economics, and the marketplace will respond enthusiastically.

For instance, PacifiCorp recently began selling two new products that help customers increase their energy efficiency in their offices, plants and buildings. Our meter Graphics program charts a customer's electricity usage in a variety of graphs, such as daily usage, peak day, and weekly load profiles. And the PacifiCorp Utility Manager program provides software to facility managers so they can quickly track their buildings' energy use, across multiple locations.

But do not expect utilities and customers to make uneconomic investments in conservation or efficiency. For example, don't expect industrial customers to be willing to pay for a refrigerator buyback that serves our residential customers.

Some communities or states, however, may desire levels of efficiency in conservation that are beyond what marketplace signals would tell us make sense. I am not opposed to those. If we are going to do that, however, we must think about who will pay for that extra level of conservation.

It's my belief that programs that are of great social benefit ought to be paid for by society, either in the form of a non-bypassable charge to all customers or by a tax. If communities decide to do that, it should be done very carefully and it should be done in a collaborative way that involves the utility, customers, regulators and special interest groups.

PacifiCorp recently put together such a group consisting of members of public and private utilities, energy efficiency and renewable energy advocates, regulators, state agency members, citizen activist groups and others.

The group has found some common ground. The majority of members agree that Oregon needs a stable way of funding conservation in the future. They believe that the way it has worked in the past – where conservation costs were rolled into rates – looks like a fairly unstable method due to the competitive pressures to keep prices low. Given this climate, the group is moving in the direction of recommending a meter charge on all customer bills as a way to fund conservation.

The group also agrees that a large part of conservation efforts should focus on market transformation activities rather than implementing costly demand-side management programs.

The group's recommendations may be incorporated into legislation that will be introduced sometime in 1997. PacifiCorp will support the meter charge, because we think it's important to have a stable source of funding that meets two conditions: it is non-bypassable, and it is competitively neutral.

At the same time, PacifiCorp will continue to support cost-effective programs and methods to achieve energy efficiency. I would like to highlight some of our efforts:

Energy FinAnswer Program: Helping customers find ways to increase their energy efficiency is both an important customer service and a valuable demand-side management resource. We believe it makes sense for the customer who receives the benefits of conservation programs to pay for the costs. That's why we developed the Energy FinAnswer Program, a nationally award-winning program that lets us work closely with our commercial and industrial customers to install conservation measures in their offices, factories and buildings. The customer pays for the costs through a monthly charge on the bill. Often, the energy savings are enough to offset most of the costs. Last year, for example, we worked with Portland Art Museum to install conservation measures that will save the museum $26,600 per year on their bill, and 652,000 kilowatts annually.

Market Transformation: PacifiCorp believes that market transformation efforts make the most sense, because they naturally move the marketplace to a higher level of conservation over time. Many of these efforts can be done collaboratively, with other utilities, state energy offices and other groups, to reduce costs and increase market penetration.

Compact Fluorescent Bulbs – This year, PacifiCorp joined with other Northwest Utilities to provide incentive payments to two national manufacturers of compact fluorescent bulbs. This program reduces the cost of the bulbs to about one-half the normal price. We believe that as customers learn how much more energy efficient these bulbs are, they will buy at higher volumes, and the price will gradually fall.

We have found that subsidizing manufacturers often makes more sense than running rebate or coupon programs, especially when the market is not very mature. Also, we've learned that rebate programs tend to have higher administrative costs.

Manufactured Homes – One of our most successful programs was a collaborative effort with Northwest Utilities and the Oregon Department of Energy to increase the energy efficiency standards of mobile homes. We provided cash incentives to manufacturers of mobile homes to build Super Good Cents homes. During the program's run, from 1991-1995, market penetration in Oregon reached 95 percent. The manufacturers found a solid market for Super Good Cents homes, so they continued to build them after the incentives were discontinued. Between August 1995 and August 1996, about 90 percent of the mobile homes built in Oregon were Super Good Cents models. It's important to note that "true" market transformation has not yet occurred, because ODOE continues to offer advertising support, and it charges manufacturers a $30 fee to certify each home. But we consider this effort well on the way to moving the marketplace toward higher energy efficiency.

Super Efficient Refrigerator Program – PacifiCorp teamed up with a group of utilities to fund a national contest challenging manufacturers to build a new CFC-free and super efficient refrigerator. Whirlpool won the Super Efficient Refrigerator program (SERP) and is selling the product in retail outlets throughout the country. PacifiCorp has helped promote SERP through public relations activities and advertising.

Oregon E-Rated Program – This fall, we began working with the Oregon Department of Energy to help promote a new program that identifies the most energy efficient and high-performance appliances available to the world. Directed at builders, this program touts these appliances as a benefit that adds value to a Super Good Cents home.

Increasing Energy Efficiency Codes – PacifiCorp is working hard to increase energy efficiency codes for buildings and homes in the seven states we serve. This is the purest form of market transformation and has a powerful impact on achieving energy efficiency in a widespread, uniform manner. We've been successful in getting stricter, more energy efficient codes in Oregon and Washington, and are pursuing initiatives in other states. To help familiarize builders and architects with the new codes, we've financed educational and training programs for them.

PacifiCorp will continue to support efforts like these, and we'll continue to have an ongoing dialogue about conservation with our customers and our regulators. In this last respect, I see the role of regulators changing as segments of the electric industry are deregulated. I think we'll see regulators focusing on facilitating solutions that meet customer needs and encouraging competitive behavior among energy providers, rather than mandating subsidies or special programs. I believe these changes will be positive, and we will find new, creative ways to deliver energy efficiency in a competitive marketplace.

ENERGY AND NATIONAL SECURITY

Caspar Weinberger, Chairman, Forbes, Inc., Served as Secretary of Defense 1981 to 1987, and Former Secretary of the Department of Health, Education and Welfare, and as Director of the Office of Management and Budget.

Moderator: It is my honor to introduce today's keynote speaker. We hope you enjoyed the two panel discussions this morning. Both this year and last year, the public policy panels showcased the differences of opinion that exist between the Clinton administration and the Republican Congress. Today's speakers also highlighted how far away we are from reaching a bipartisan consensus on our nation's energy policies. But while we argue over details, we should not lose sight of the big picture, and there are very few people more qualified to present that picture than today's keynote speaker, Caspar Weinberger. For more than 25 years, he has viewed the world from a number of unique vantage points. His impact on the world scene has been tremendous.

He first came to Washington during the Nixon and Ford administrations, where he served as Chairman of the Federal Trade Commission, as Director of the Office of Management and Budget, and as Secretary of Health, Education & Welfare. In each of these positions, he has served our country with distinction. Had he done nothing more his legacy would have been substantial. But Caspar Weinberger's vision got only bigger as he tackled the international scene. He served as Secretary of Defense for President Reagan from January 1981 until November 1987. In that capacity he oversaw the rebuilding of our national defense. As Secretary of Defense, he was influential in making the far-reaching policy decisions that brought about the collapse of the Soviet Union and the successful conclusion of the Cold War. As a result of his influence, the US was able to build a high-tech armed forces that defeated Iraq with a minimum of Allied bloodshed. Our armed forces are the strategic and technological leaders of the world and the US is the only superpower, due in large part to his efforts.

His legacy is acknowledged throughout the world. He has received honors from countries ranging from Great Britain to Thailand. After leaving the Defense Department, Caspar Weinberger was a member of the Foreign Intelligence Advisory Board. He became a publisher of Forbes, the biweekly business magazine in 1989, and chairman of Forbes in May in 1993. He remains an active witness to the world scene, and his advice and counsel are still sought by political leaders. We are fortunate today that he will share his views with us. So, ladies and gentlemen, please welcome Mr. Caspar Weinberger.

Caspar Weinberger: It is very nice to be here and talking about some of the more important issues that are going to be facing the world for quite a long time. And I think it is important we realize just how vital they are for our future. You heard a lot of very important and expert technical matters discussed today by people who really know what they are talking about. I've been asked to step back and review energy policy in the context of national security. That really is another way of saying I'm clearly not an expert, and I think the role here that I have to play is essentially that of concerned citizen. And I think there is a great deal to be concerned about.

One aspect of the nation's energy policy that doesn't really receive much attention until we have to call out our armed forces to secure our oil supplies, is the continuing basic threat we have when we have such a very heavy dependence on foreign oil. I don't point out particularly any one threat from any one country or anything of that kind, but when we have to have such a very large amount of reliance on different foreign countries, particularly from the Middle East which is a very unstable and volatile area, it is important to realize that we are still importing right now. We are now importing again about 50% of the 17 1/2 million barrels of oil that we use every single day. By the year 2010 it has been estimated that about 70% of our oil will have to be imported and most of that from the Persian Gulf, if the current policies continue and unless we are able to make some spectacular and certainly unplanned for new discoveries.

Our trade deficit in oil is expected to double which will mean we'll reach about $100 billion and that's in the same time frame, by the end of the year 2010. And imports at the level of 70%

would be too much and I think it would compromise both our security, our economic growth, because when you have to transfer $100 billion of your treasury to other countries, it has to affect adversely the economies of our own country.

I think everyone here is familiar with the general energy situation. Even though we account for 30% of the consumption of the world's petroleum, we only have about 3% of the world's proven reserves. And falling energy prices over time have now actually increased the consumption even as our domestic production has dropped. I certainly don't favor any artificial measures such as command control economies or high energy taxes or anything of that kind which would increase energy costs in a deliberate attempt to drive down the usage. Because I think what you do then is affect adversely economic growth, and that I think hurts people more than anything else.

At the same time I do oppose some of these non-market measures that have been used to lower energy prices. In what was, I think, a spectacular vote-getting attempt President Clinton recently manipulated the gas prices by releasing some oil from the Strategic Petroleum Reserve. This is not going to be anything that is going to solve the problem or even make more than a small dent in it. Actually, it can exacerbate the problems in the long run because his manipulation doesn't solve the problems that the oil industry is facing, which is a very high level of oil imports. And it has caused the closure, indeed, of many domestic refineries, and put many Americans out of work. Finally, merely allowing a little bit out of the Strategic Petroleum Reserve is not going to do anything more than perhaps endanger some of the supply we may urgently need at crisis time, which is of course the reason why the SPR was created.

There is nothing particularly unique about this to the US. What is happening here is happening to other nations throughout the world. Energy consumption in skyrocketing and energy growth rates are higher in many developing countries right now than in the US. The World Bank has estimated that by the year 2010 there will be 800 million vehicles in the world. I was in Brazil recently on Forbes matters and in Brazil alone there are several new automobile manufacturing plants that will come on-line in the next two to five years. They are, of course, delighted. They are trying to get them. It helps their economy enormously with their unemployment problems, but of course that will mean that Brazil's electric and oil use will have to increase again. They increased 8% last year which was double the increase of their GNP. As in other developing countries, surges of economic growth and increased industrialization means, of course, increased energy usage.

In the growth of the Pacific Rim countries, which are of course are the most astonishing economic engines in the world at the moment, if that continues at the present rate, their energy demands will very soon really exceed those of the US. Pacific Rim countries are forecasted to become the biggest energy-consuming nation before the year 2000. That used to sound such a long time away. It's just a little over three years now. Their oil needs, of course, if things go the way they are now, will again, have to be met largely from the Middle East. Perhaps as much as 90% of those increased needs for the Pacific Rim will have to come from the Middle East as, indeed, is much of the oil now that Japan needs for their heavily industrialized society.

So even more disturbing, I think, is the prospect of a disruption then of the supply of oil. Tremendous growth and travel, and little or no emphasis on energy efficiency in the last few years, increasing environmentalism helping to cause declining domestic oil production, all have made the U.S.'s self-sufficiency in oil, something we used to talk about 25 years ago, virtually impossible. I think we will always have to import some oil. The question is how much and from where, so it's not really a question of whether there will be an increase in imports – certainly that, but now it's actually trying to decrease the amount that we import. If there were any interruptions in production, any overseas activities that resulted in war or closure of the sea lanes or anything like that, the economy here could be brought to a total standstill in something under 90 days. That's what the Strategic Petroleum Reserve was supposed to avoid and 90 days is not a long time in the world of today.

If another major energy crisis were to occur, that would happen very, very quickly and we would have to ask ourselves, as we do now, are we ready for that. And clearly the answer is no. In this world at a minimum the US will face ever increasing economic costs. The OPEC nations have made it quite clear thus far that they don't intend to increase production in a drastic way over the long-term. That, of course, has, and it is intended to have, the effect of driving all prices up. We spend about a billion dollars a week now to pay for oil imports. That represents about half of our trade deficit at the present time. Trade deficits can be argued about one way or another, whether they are good or bad. Frequently they aren't all that bad because they generally mean we are also selling some things abroad and that it's not all that difficult.

But when it's concentrated in one area, such as energy, which we have to have, and since we have largely deprived ourselves of other sources of energy by our environmental policies, particularly with respect to nuclear plants, I think it is clear that as demand increases and the bill for importing oil continues to grow and no country can really continue to afford the increasing transfer of resources from ourselves to any other country without causing very serious harm to our economy. You can't really make up for it by other areas of production, or other areas of the economy, when such a huge amount has to go to just for oil and for energy.

Obviously we need to look at all the political and security implications of the world energy supply and we have to create our energy policies with that understanding of what's happening in the world clearly in mind. This is not just a product that you can import or export or consider in any sense as a part from the world's political problems, because this is imminently tied up with, as we saw in the case of the Gulf War, securing our oil supply practically overnight, we found ourselves in a major war, not just to deter the aggression against Kuwait, but to secure a reliable US supply of oil at reasonable prices. Because there is no question now, and certainly not in my mind on the basis of intelligence reports and other things we had at that time, that Iraq's intention was not to stop at Kuwait but to go down into the Kingdom, and then into the Gulf, at which point they would have controlled over 70% of the world's known oil reserves. And that's not a situation we could have lived with.

That is the kind of thing that can happen very, very quickly. We need to factor in all those potentialities for instability in the Middle East which from time to time seem to be less and at other times seems to grow. At the moment we are in a period where the instability seems less likely to be removed in the immediate future than we had reason to hope for a short time ago. We had all hoped that that is not now the correct prediction. In fact, it's not a prediction at all, really. It is just an example of how quickly facts can change from a situation that could produce more stability, which is the sort of thing that we need, to one of greater instability.

As happened before, just a few years ago, with this enormous transfer of wealth from the oil consuming nations to the oil producing nations of the Middle East, they could, again, triple their resources and their revenues and with that gain, not only gain considerable influence but purchase a large number of sophisticated and particularly unpleasant weapons. That would make those nations even more of a threat than some of them are today. And certain Iran and Iraq remain a threat and will remain so, I'm sure, as long as the regimes that they have continue in office.

And so we have to our alliances, we have to look to our friendships, we have to realize that we can't do any of these things alone, that we need all kinds of alliances and friendships. And finally we have to realize that no matter good our alliances are, or how good are relationships are, things could change very, very rapidly. I always recall that for many years we had a very close working relationship with Iran. We lost it ourselves, practically overnight, because we didn't continue our support of the Shah, and that was because many liberals felt that the Shah was too repressive. And that is a very serious example of what happens when you don't consider the alternative. Because the alternative is, of course, the most repressive government since the Middle Ages and we brought a lot of that on ourselves by not recognizing that it was necessary to continue our support of as good a friend as we had in the Shah at that time and how vital it was to do that.

Well, what should really be the basis of this international energy policy? I think we have to take a few steps very quickly and immediately. One of the first is to: reduce the level of imports and, of course, reduce our own usage of oil. We are a mature oil-producing nation now and we know for the most part, where we think most of the oil is, here and offshore. But we are never certain of that. There is right now, as many of you know, a very large new potential development coming off Bag, California. That could help things out enormously if it comes in. We don't know whether it will or not.

But we do know that for one thing, we are not recovering all the oil that is and remains in the ground after conventional recoveries. A lot of work has been done, and a lot of work is being done on new methods of going down far deeper than we did before, bringing up far more oil and still doing it within economically feasible terms. There are some estimates that indicate that 300 billion barrels of domestic oil are simply bypassed because we are not using the advance technologies that are now available to us to bring that to the surface. And that many of these wells are being abandoned because the operating costs are high, and many are being abandoned because the price of this oil is felt not to be economically feasible. The answer is not to raise the price of oil. The answer is to produce some kind of inducement for drilling techniques that employ these latest technologies. We used to talk about offset drilling in California years ago. It was considered rather a problem and we had all kinds of legal discussion with it. It never amounted to more than going down and then drilling to one side. Now they can go down and go to several sides and in effect, follow a stream all the way down and the same single drilling platform. So there all kinds of things that are being developed that should be used.

And we have to look at new energy sources. It is, I think, very unfortunate that we and so much of the world have given up on the alternative source of nuclear power, particularly the development of fusion which would be a lot cleaner and perhaps safer, than the fissionable products we use now. I don't think you can build a nuclear power plant in the country now. It took about 14 years to bring the last US nuclear plant on line. Seven of those years were spent in dealing with the regulatory process and trying to go through the environmental impact statements, and getting them approved, even though once an impact statement is approved by the courts. A report simply goes on a shelf. But it is a prime method of delaying construction, which was its original intent. They have succeeded very well in that.

But you can't keep capital tied up for 14 years while you work through a regulatory process, and so people simply are not building nuclear plants. But we could spend a little more research money on fusion technologies and we would get around some of the problems we've had and we could get nuclear power back as an important energy source.

Also, there are other sources of energy that we have not been exploring very thoroughly. I think we have to encourage energy efficiency too at the same time. We have to encourage the ways in which we can reduce our usage and that means a number of things. It means conducting energy education, re-educating consumers on the technologies that exist to conserve energy. Things that you've been talking about this morning, and increasing our investment in research, to continue our leadership in energy saving technologies. These, again, are part of our national security and it is essential that we do them.

I think it is possible to do a great deal of this kind of thing without in any sense reducing our quality of life. I think that we have to recognize that we have a great many things that are prevented when this sort of a crisis continues – things such as going back to a rural life and not bothering with growth any more, or with cities, and discouraging anything that involves growth because it involves more use of energy. None of that is either necessary or sensible. But I think that what we have to do it realize that there are things that can be done and inducements that can be offered in the form of tax benefits and other things, that will encourage the greater reduction of usage.

In the DOD when I was there and talking to some of the people involved in it this morning, we were able to make very substantial savings, up to billions of dollars, simply by looking through the Department, and a number of things could be done to reduce energy consumption, such as having more meetings by teleconferencing instead of requiring travel, and all of those sorts of things. And if we put our minds to it, we've been able to solve so many other problems, I think we can do it without in any sense reducing the quality of life or asking people to give up driving or any of those measures which we've talked about so confidently a few years ago as being the only way to solve the problem.

Energy efficiency has the support of the whole business community. It is not a problem in which the business community is opposed. Quite the contrary. The Business Council for a Sustainable Energy Future has more than 100 firms now, ranging from the Forbes 500 companies to small entrepreneurs have said that energy efficiency and renewable energy programs produce new and improved products that can create really tens of thousands of new jobs, just in the production of those energy saving devices. And that strengthens our competitiveness abroad and through partnerships with local private sector companies, I think we can reduce the need for any additional federal resources.

I think we need to understand that energy efficiency does not necessarily need to be a hardship. And that is important to bear in mind, because we really don't have to do what is neither desirable or possible, and that is asking people to change their major lifestyle. We have achieved a better lifestyle and quality of life and living, for a larger number of people, than any other country in any other civilization in the world's history. There is no need to say that all that has to be abandoned because we have problems in supplying the energy necessary to maintain it. The challenge ought to be to maintain it and to do it with using less energy. And that I think is perfectly within our compass and within the technologies that are here.

Through incentives, we have done something. We could do more to encourage both companies and individuals to be more energy efficient. We made some substantial progress on this in the middle of the last oil crisis. But as that subsided, and as the need for it was perceived to be no longer there, we pretty much reverted to older ways.

Government leadership can come in a number of ways and it doesn't mean increased spending. That is, the normal immediate reaction when you have a problem is to turn to the government and ask them to spend more money. And, as usual, that is not going to have any better effect than most of the other government programs that have been born with that kind of a concept

I don't think you are going to have any particularly desirable results with government mandates. A lot of people were talking a little while ago about government mandates that are designed to produce zero growth and a simpler rural life, with less machinery and fewer comforts. I think most people who experienced those don't really want to try them again, no matter how attractive they might sound from a distance.

I think government can be the role model. They can certainly do a lot of public education. They can do a lot within their own house as we did in the Defense Department in reducing the usage of energy. And there are a lot of other activities such as this energy efficiency forum in which you have been discussing these matters so effectively today.

When we reduced energy in the DOD, we didn't have to focus so much on saving more money, because energy conservation did have those results. They did save money. But we also wanted to make sure that the government of the US and the defense of the US was not increasingly dependent on a supply from a source that neither at that time nor now, could be called completely and wholly reliable, particularly when our governmental polices have not been such as to improve the reliability of some of those sources by our other actions.

One thing we tried to do in Department of Defense was to develop a single fuel on the battlefield. That significantly reduced the supply requirements of the front line. We tried our best to get rid of the requirement that had been imposed, because of widely understood Congressional reasons that required the US forces in Europe <u>not</u> to use European coal but to import coal from Pennsylvania to Europe to use there. Then we found that some of the energy plants in Europe couldn't use our Pennsylvania coal. Nevertheless, we still have to do that because this was good for a congressional district or two.

There are things like that that you can do that can avoid not only excess use but a fairly useless use of additional energy. We had the National Energy Awareness Proclamations for the people at the DOE and I think made some progress toward improving the acquisition of petroleum and streamlining legislation, and that saved both energy and taxpayer dollars.

There is no doubt internationally, as was said in that very nice introduction, a number of nations that look to us as a mentor and regard us as the only military superpower and we have to fulfill to that role, too, and guide the way to a sustainable energy future for all friendly countries. One way to do this would be to encourage businesses to share information and thereby open some markets. And we can encourage a lot more <u>cooperation</u> both internationally and domestically. If we are going to do that we have to assure the companies that were asked to do that, that they are not going to be subjected to later prosecutions for antitrust violations, which would be another example of how one hand of the government doesn't really pay much attention to what the other hand is doing.

Other countries have things we can learn from. Taiwan has a huge infrastructure development plant for the next 10 years and they are going to need a lot of energy. They also need development to satisfy their enormously successful and expanding economy. We can certainly have a role there and they are eager for us to do it, to pass on our experience in how a lot of things could be done with less than what seemed to be the initial energy requirements. So I think we could do quite a lot with international exchanges with the free exchange of information and experiences we have had.

All this really comes back to is the basic point that I tried to make in the beginning. I think we do have to draw a line in the sand that says basically 50% oil imports and no more. And we should have a lot of information as to when we are getting to that point and when we're crossing it. I think very few people realize that we are now at 70% and I think that is infinitely too much. We are always going to have to import some oil, as I said earlier. We will weaken ourselves if we don't recognize that fact. But we can certainly reduce the amount we have to import and we can take a far more active and informed interest in the politics and policies of the governments from which we have to import oil. That should be another part of the comprehensive overall integrated approach to the whole problem. We have to think what we can do to help make some of our sources more stable.

We need to emphasize that growth can and must continue. We need to emphasize that by reducing the reliance on imported oil we help our security, and we need to encourage the production of more domestic oil, by all kinds of tax and other incentives, and particularly the ability to recover domestically, much of the oil that we have simply walked away from saying it is too expensive, or not as profitable as taking oil from higher up in the well.

And, of course, through education we can certainly help and encourage business upgrade building systems and retrofit old products and controls with these new energy saving technologies. We can't go back to the policy in which the model is less growth, less development, or says let's return to a nice, easy and comfortable rural life. We can, I think, obtain energy efficiency through technologies without impeding our national growth or our energy levels. And I think that has to be done and it has to be done quickly.

One of the things that is best to do is what you have been doing this morning. And we are in debt to the people who organized and conducted this forum, because this is the kind of thing that can provide more and more people with the knowledge of what has to be done. The fact that it can be done and it could be done without seriously interfering with the basic quality of life of many people. So I think the educational aspect of all of this is enormously important and I congratulate all of you on the very fine work that has been done this morning, and I can only urge and request that you continue as missionaries for this work, this awareness, and this additional education in the future. Thank you very much indeed.

Moderator: Thank you, Mr. Secretary. Your comments certainly will serve as a framework for the rest of us here today at the rest of the forum. The Secretary has agreed to answer a few questions, and if anybody filled any of the cards out we can read them and see if you can answer those.

Q: You mentioned your support for fusion nuclear research. Do you also support federal research on energy efficiency technologies?

Weinberger: Yes. I think this can certainly be encouraged. We have already put a fair amount of money into fusion research, but we have never really had, I think, the commitment really to try and develop usable energy from fusion. And I think that would take additional government funding and I could make some recommendations as to what could be cut out to find that funding.

I think the government encouragement of educational programs can be done with very little money, but they need a lot of leadership qualities, and with the encouragement of forums such as this and with the number of inducements for corporate achievements, individual achievements, things of that kind, that can be basically beneficial and I hope that would be done as part of a coordinated policy, that is one part of a number of different things that need to be done, basically, all at once.

Q: Please comment on the delivery of oil from the former Soviet Union, especially the non-Russian republics and its implication for the US.

Weinberger: Well, here again you have an attempt to estimate what's going to happen in foreign countries, and whether or not we would be safe in relying on a lot of that. When I was in office, we had the big question of the Russian pipelines that were being built at that time and a lot of European countries worried about prices from other countries, then were trying to sign up and take oil and gas from these pipelines. We urged that they not do so, on the ground that all of the valves were in the Soviet Union and they might wake up some morning without having anything come out their end. That does seem to point up the problem that we have now. Now we have a different government in Russia. We aren't quite sure which one it is yet, but it is going to be different. We also have a lot of attempts in Russia and some resistance in other republics to bring back the old Soviet Union or something that looks like it. These are, again, reasons why the campaign and the energy policy needs to be coordinated with foreign policy and needs to be based upon what's happening in other countries and whether or not that is going to be a safe or a desirable source of alternate energies or alternate supplies of energies.

I think if we had a Russia we were sure was going to be friendly, and we're sure that it was going to want to help the West, and we had a lot of doubts about that in the campaign that just closed because several of the candidates were mouthing all the anti-Western slogans. They didn't get elected but they got a lot of votes. We don't know what the final results will be. But when you rely on other areas for your energy supplies to the extent we are doing now, and when you don't have any policy that coordinates the policy of the government, even our defense and security policies, with the countries as suppliers in mind, then I think you risk your national security in a way that is very undesirable.

CONCLUSION

Barry Worthington, Executive Director, United States Energy Association.

In reviewing both the success and the substance of the first seven forums, while simultaneously planning for the eighth forum in 1997, one is struck by several key phenomena that are evident year after year:

— The spirit of bipartisanship.

— The commonality of interest.

— The sharpness of debate over polarizing issues (and perhaps most striking).

— The complete consensus that evolves support efforts to improve energy efficiency.

Few issues attract equal support from Republicans and Democrats. Perhaps fewer still between the administration of George Bush and Bill Clinton. Yet, both major political parties and both recent U.S. administrations recognize the explicit value of energy efficiency.

How many issues exist that draw support from environmentalists and energy producers? Both are on record at these forums expressing strong support for improving our nation's energy efficiency.

Concerns over the potential climatic effects of increasing greenhouse gas emissions, and the impact of electric utility industry restructuring, have been perhaps the most hotly contested energy issues thus far in the 1990's. Both of these polarizing issues have been boldly and frankly addressed at this unique series of forums. Republicans or Democrats, Bushites or Clintonites, business leaders or environmentalists, all agree on at least one thing: energy efficiency is an important issue. The United States must and will do more.

Bipartisan Venue

While no one disputes the need for a consistent and sound national energy policy, translating this consensus into legislative and administrative action is much more difficult. Since the first forum was held in 1989, we have had three Secretaries of Energy, and the most turnover in the U.S. Congress in decades. The forum serves a crucial role of providing perhaps the only continuing bipartisan venue to focus the attention of both government and business leaders on this issue.

Serving as a bipartisan forum has been critical. Pressures for political posturing, deep philosophical differences, and the need to focus on other major issues, often lead to rancor and divisiveness in Washington. Fulfilling the commitment inherent in the forum's motto, Pledged to United States Leadership in Energy Efficiency is only possible due to our ability to attract the highest level of participation from both government and the corporate community.

Attendees at the forum have discovered it to be an outstanding networking opportunity. Most participants attend year after year. Hundreds, if not thousands, of business relationships have evolved. Perhaps the most compelling example is the opportunity to meet the international delegations. Senior energy efficiency policy-makers from more than one dozen countries typically attend each forum. These international delegates are sponsored by the U.S. Agency for International Development (USAID). A heartfelt gratitude from all interested in encouraging global energy efficiency is extended to Robert Ichord and Jefferson Seabright of USAID for their role in extending on a global scale, the collaborative process inherent in the USEA/Johnson Controls Energy Efficiency Forum.

Like many individuals and organizations that have become forum supporters, Jeff Seabright and Bob Ichord are on the front lines every day dealing with implementation tactics and budget battles to push for adoption of energy efficiency strategies and policies. We have good reason to believe the other 300-plus attendees each year are serving as yeomen in their own right, within their own organization and their own sphere of influence.

World Energy Congress

The message of the forum will be elevated to a truly global plateau when the United States hosts the 17th Congress of the World Energy Council in Houston, Texas, in September 1998. Four heads of state, 125 energy ministers, 2,000 corporate officers and 5,000 additional experts, specialists and journalists will descend on Houston – for what some have in a modest overstatement called the Triennial Olympics of the Energy Industry – and discuss, debate and, hopefully, reach a consensus on several of the exact issues that have been a focus of the USEA/Johnson Controls annual forum.

Our work does not stop with either the Eighth Annual USEA/Johnson Controls Energy Efficiency Forum in 1997, or the 17th Congress of the World Energy Council in 1998. Rather, these events are only the beginning of our challenge.

In an era of stable or even falling energy prices that most likely will be accompanied by abundant supplies and no price or supply disruptions, one can fairly ask the question, why should anyone care about energy efficiency? After all, the U.S. Department of Energy is projecting oil prices in 2015 to be less than the market price in December 1996. Both government and industry forecasts indicate adequate fossil fuel resources to meet demand well into the next century. This rosy scenario suggests that we can use all the energy we can pay for, so why worry? The reality is if we know one thing about forecasts, it is that they are almost always wrong. A bit of hedging against inaccurate forecasts is only prudent.

Raising Energy Questions

The relevant question is, "To do or not to do – energy efficiency and at what cost?" This cannot be answered without raising additional questions: What is our exposure to increasing imports? What is our exposure to unpredictable global and domestic events? What is our exposure to price spikes or supply disruptions? What is our exposure to known or unknown environmental problems including the impact of increasing greenhouse gas emissions? Of course, the only legitimate answer is, "No one really knows." But, we do know that energy efficiency provides a strategy to hedge our bets.

The questions, however, don't end there. What is the economic cost of premature or unnecessary action to reduce greenhouse gas emissions? What is the economic and environmental cost of mandating alternative energy sources? What is the economic and societal costs of job elimination if we shut down domestic industries needlessly?

No magic answer exists to solve the problems of the world, either, but energy efficiency, based on economic rationality, truly is a no-lose proposition. Energy efficiency may improve corporate profitability, reduce government expenditures, create jobs, reduce import dependency, extend the availability of traditional resources and, should there be a need or value, energy efficiency is certain to be the most cost effective mechanism and most market-based mechanism to reduce greenhouse gas emissions.

Only The Beginning

By now, it should be clear that given the outstanding contributions of the USEA/Johnson Controls Energy Efficiency Forum, our work is nevertheless only the beginning. The issues raised at the first seven forums – highlighted in this anthology and hopefully adequately summarized with a sense of drawing out key conclusions in this, the final chapter – clearly will provide fodder for future forums and for our national agenda.

I predict that as we enter the next millennium, this event will be the standard by which other forums are measured. I cannot conclude my contribution without strongly acknowledging the role that Doug Decker has played in this forum. Doug is the creator, the inventor and the inspiration behind the USEA/Johnson Controls Annual Energy Efficiency Forum.

As USEA looks forward to the next century, we are most pleased to continue this focus on energy efficiency and to recognize our partner in providing the only nonpartisan, non-political, objective, government/industry forum to mutually explore energy efficiency as a national priority.

Rep. Scott Klug (R-Wisconsin) voices the concerns of the freshman Congressional delegation in 1996.

At the 1996 forum, Christin Ervin, Assistant Secretary, Energy Efficiency and Renewable Energy, Department of Energy, defends the expenditure of scarce funds on energy efficiency projects with quick payback and multiple benefits.

Former Defense Secretary Caspar Weinberger (left) discusses current Defense Department energy strategies in 1996 with Millard Carr, Director, Energy & Engineerings, Office of the Assistant Secretary of Defense.

In 1996, Jefferson Seabright, director of USAID's Office of Energy, Environment and Technology, offered several ideas about promoting and paying for energy efficiency abroad.

USEA Executive Director Barry Worthington concludes the 1996 Energy Efficiency Forum with a look toward the 1998 World Energy Congress in Houston.

Rep. Barbara Vucanovich (R-Nevada) receives the 1996 Energy Leadership Award from Johnson Controls Executive Vice President Joe Lewis.

Johnson Controls Vice President Brian Stark presents the 1996 Energy Leadership Award to former Defense Secretary Caspar Weinberger.

Appendix

Energy Leadership Awards

Each year, members of the Johnson Controls/USEA Energy Efficiency Forum honors individuals by recognizing their roles in promoting energy efficiency on the state, national or international level while in public service.

The 15 people who have won the awards over the years represent an interesting mix of Democrats and Republicans, politicians and public servants. In their acceptance speeches, the winners usually thanked members of their staffs and other personal friends and relatives. These remarks at conclusion of the 1995 Awards Presentation by DOE's John Millhone, himself the 1994 recipient of the 1994 Public Service Award, seem to most thoughtfully express the feelings of the award winners through the years.

One of the characteristics that I think identifies those who are recognized by these awards is the durability that they have had to work successfully in a complicated area time and time again. I think that energy efficiency is one of those few areas that lends itself to that kind of ability to redefine what you are going to be gaining.

If you are interested in saving money – saving money through energy efficiency is one of the most cost-effective, significant, quick-return ways to get things done. At other times, if you are interested in doing something that benefits the environment – energy efficiency, by reducing greenhouse gas emissions, is one of the most significant things you can do. If you are interested in national security – we've got the answer: reduce oil imports.

If you're interested in ozone depletion problems – as long as you have to have CFCs, why don't you replace them with new equipment that is energy efficient? If you are concerned about the competitiveness of the U.S. internationally – one of the most important things is to reduced the energy component of the product that we export, and we do it through increased energy efficiency.

The people who are recognized with awards stand for something. They are an inspiration for some of us in the public sector who have been trying to figure out ways we can work with Congress to be effective. It's sort of a merger of the ability to see all of the benefits that come from what we are doing and then figuring out how to describe what we want to accomplish.

In a way, that is consistent with that changing definition of what's important, and at the same time stand for something that is significant that pulls these different things. That, I think, is the quality of what has been most successful in energy efficiency activity.

AWARD WINNERS

Year	Recipient	Award
1991	Congressman Les Aspin	Congressional
1992	Senator John Glenn	Congressional
1992	Congressman David Obey	International
1993	Dr. James Sullivan (USAID)	International
1993	Governor Tommy Thompson	Governor's
1994	Congressman Philip R. Sharp	Congressional
1994	Governor William Donald Schaefer	Governor's
1994	John P. Millhone (DOE)	Public Service
1995	Robert F. Ichord, Jr. (USAID)	International
1995	William K. Reilly (EPA)	Public Service
1995	Congressman Mike Synar	Congressional
1996	Former Secretary of Defense Caspar W. Weinberger	International
1996	Secretary of Energy Hazel O'Leary	Public Service
1996	Congresswoman Barbara Vucanovich	Congressional

AWARD CRITERIA

Criteria for the four awards (each of which is not necessarily presented every year) are as follows:

- The Congressional Award recognizes a current or former member of Congress, or congressional staff person who has contributed to legislative achievement so encourage energy efficiency gains, particularly in federal government activities.
- The Public Service Award recognizes a current or former federal or state government official (non-elected) who has motivated action to addressing energy efficiency in government operations and has served as an advocate of energy efficiency.
- The International Award recognizes a current or former public official who has encouraged energy efficiency internationally.
- The Governor's Award recognizes and individual who has contributed to energy efficiency at the state government level. The recipient may be a current or former governor, agency executive, commission member or other state official.

Nominees for all awards must be U.S. citizens and must be current or former public servants. Each nominee is generally recognized among peers for his or her achievements in energy efficiency.

Nominations are solicited from USEA members, acknowledged energy efficiency specialists and peer groups, and were submitted to USEA for review by its Energy Efficiency Committee. Final selections were made by the USEA Energy Efficiency Forum Committee.

ENERGY EFFICIENCY FORUM HONORARY CO-HOSTS (1991-1996)

GOVERNOR GEORGE ALLEN
GOVERNOR EVAN BAYH
GOVERNOR TERRY BRANSTAD
GOVERNOR MEL CARNAHAN
GOVERNOR JOHN ENGLER
GOVERNOR JIM GERINGER
GOVERNOR WILLIAM JANKLOW
GOVERNOR BOB MILLER
GOVERNOR BEN NELSON
GOVERNOR GEORGE PATAKI
GOVERNOR ROY ROMER
GOVERNOR WILLIAM DONALD SCHAEFER
GOVERNOR DON SUNQUIST
GOVERNOR TOMMY THOMPSON
GOVERNOR GEORGE VOINOVICH
GOVERNOR WILLIAM WELD
GOVERNOR JOHN WAIHEE
GOVERNOR LOWELL WEICKER, JR.
GOVERNOR PETE WILSON

SENATOR SPENCER ABRAHAM
SENATOR JEFF BINGAMAN
SENATOR DAVID BOREN
SENATOR JOHN BREAUX
SENATOR THAD COCHRAN
SENATOR PAUL COVERDELL
SENATOR ALFONSE D'AMATO
SENATOR DENNIS DECONCINI
SENATOR PETE DOMENICI
SENATOR RUSSELL FEINGOLD
SENATOR BILL FRIST
SENATOR JOHN GLENN
SENATOR BENNETT JOHNSTON
SENATOR HERB KOHL
SENATOR TRENT LOTT
SENATOR HARLAN MATHEWS
SENATOR DON NICKLES
SENATOR RICK SANTORUM
SENATOR JIM SASSER
SENATOR FRED THOMPSON
SENATOR JOHN WARNER
SENATOR PAUL WELLSTONE

ENERGY EFFICIENCY FORUM HONORARY CO-HOSTS (Cont.) (1991-1996)

CONGRESSMAN PETER BARCA
CONGRESSMAN THOMAS BARRETT
CONGRESSMAN HERBERT BATEMAN
CONGRESSWOMAN HELEN DELICH BENTLEY
CONGRESSMAN HOWARD BERMAN
CONGRESSMAN BILL BILBRAY
CONGRESSMAN JAMES BILBRAY
CONGRESSMAN THOMAS BLILEY
CONGRESSMAN SHERWOOD BOEHLERT
CONGRESSMAN GEORGE BROWN
CONGRESSMAN ED BRYANT
CONGRESSMAN SONNY CALLAHAN
CONGRESSMAN BENJAMIN CARDIN
CONGRESSMAN JIM CHAPMAN
CONGRESSMAN DICK CHRYSLER
CONGRESSMAN WILLIAM CLINGER
CONGRESSMAN RONALD COLEMAN
CONGRESSMAN JIM COOPER
CONGRESSMAN SAM COPPERSMITH
CONGRESSMAN NATHAN DEAL
CONGRESSMAN NORMAN DICKS
CONGRESSMAN ROBERT EHRLICH
CONGRESSMAN JOHN ENSIGN
CONGRESSMAN SAM FARR
CONGRESSMAN VIC FAZIO
CONGRESSMAN BOB FILNER
CONGRESSMAN ERIC FINGERHUT
CONGRESSWOMAN ELIZABETH FURSE
CONGRESSMAN PETE GEREN
CONGRESSMAN STEVE GUNDERSON
CONGRESSMAN GIL GUTKNECHT
CONGRESSMAN LEE HAMILTON
CONGRESSMAN J.D. HAYWORTH
CONGRESSMAN GEORGE HOCHBRUECKNER
CONGRESSMAN WILLIAM HUGHES
CONGRESSMAN SAM JOHNSON
CONGRESSWOMAN MARCY KAPTUR
CONGRESSMAN JOSEPH KENNEDY
CONGRESSMAN JACK KINGSTON
CONGRESSMAN GERALD KLECZKA
CONGRESSMAN HERBERT KLEIN
CONGRESSMAN SCOTT KLUG
CONGRESSMAN MIKE KOPETSKI
CONGRESSMAN JOE KNOLLENBERG
CONGRESSMAN MIKE KREIDLER
CONGRESSMAN RON LEWIS
CONGRESSMAN BOB LIVINGSTON
CONGRESSWOMAN MARILYN LLOYD
CONGRESSMAN RON MACHTLEY

ENERGY EFFICIENCY FORUM
HONORARY CO-HOSTS (Cont.)
(1991-1996)

CONGRESSMAN THOMAS MANTON
CONGRESSMAN EDWARD MARKEY
CONGRESSMAN MATTHEW MARTINEZ
CONGRESSMAN ROBERT MATSUI
CONGRESSMAN DAVE MCCURDY
CONGRESSMAN JIM MCDERMOTT
CONGRESSMAN DAVID MCINTOSH
CONGRESSMAN MICHAEL MCNULTY
CONGRESSWOMAN CARRIE MEEK
CONGRESSMAN JOHN MICA
CONGRESSMAN DAVID MINGE
CONGRESSWOMAN PATSY MINK
CONGRESSMAN CARLOS MOORHEAD
CONGRESSMAN JAMES MORAN
CONGRESSMAN MARK NEUMANN
CONGRESSMAN CHARLIE NORWOOD
CONGRESSMAN JAMES OBERSTAR
CONGRESSMAN DAVID OBEY
CONGRESSMAN JOHN OLVER
CONGRESSMAN COLLIN PETERSON
CONGRESSMAN THOMAS PETRI
CONGRESSMAN DAVID PRICE
CONGRESSMAN NICK RAHALL
CONGRESSMAN BILL RICHARDSON
CONGRESSMAN FRANK RIGGS
CONGRESSMAN TOBY ROTH
CONGRESSMAN DAN SCHAEFER
CONGRESSWOMAN PATRICIA SCHROEDER
CONGRESSMAN JIM SENSENBRENNER
CONGRESSMAN PHIL SHARP
CONGRESSMAN CHRISTOPHER SHAYS
CONGRESSWOMAN KAREN SHEPARD
CONGRESSMAN DAVID SKAGGS
CONGRESSMAN JIM SLATTERY
CONGRESSWOMAN LOUISE SLAUGHTER
CONGRESSMAN NICK SMITH
CONGRESSMAN LOUIS STOKES
CONGRESSMAN GERRY STUDDS
CONGRESSMAN MIKE SYNAR
CONGRESSWOMAN KAREN THURMAN
CONGRESSMAN EDOLPHUS TOWNS
CONGRESSWOMAN JOLENE UNSOELD
CONGRESSMAN BRUCE VENTO
CONGRESSWOMAN BARBARA VUCANOVICH
CONGRESSMAN HENRY WAXMAN
CONGRESSMAN DAVE WELDON
CONGRESSMAN ALAN WHEAT
CONGRESSMAN ED WHITFIELD
CONGRESSMAN DON YOUNG